Encyclopedia of Earthquake Research and Analysis: Seismological Advances

Volume IV

Encyclopedia of Earthquake Research and Analysis: Seismological Advances Volume IV

Edited by **Daniel Galea**

New York

Published by Callisto Reference,
106 Park Avenue, Suite 200,
New York, NY 10016, USA
www.callistoreference.com

**Encyclopedia of Earthquake Research and
Analysis: Seismological Advances
Volume IV**
Edited by Daniel Galea

International Standard Book Number: 978-1-63239-237-4 (Hardback)

Printed in the United States of America.

Contents

Preface

The aim of this book is to educate readers regarding the seismological developments in the field of earthquake research and analysis. The study of earthquakes involves science, technology, infrastructure and engineering in an effort to reduce human and material loss when their occurrence is inevitable. This book discusses various aspects of earthquake research and analysis, from theoretical advances to practical applications. It discusses ground tremor studies and seismic site characterization, with regard to reduction of risk from earthquake and ensuring the safety of infrastructure under earthquake loading. The objective of this book is to stimulate discussions and research for improving hazard assessments, distribution of earthquake engineering data and most importantly, the seismic provisions of building codes.

The information shared in this book is based on empirical researches made by veterans in this field of study. The elaborative information provided in this book will help the readers further their scope of knowledge leading to advancements in this field.

Finally, I would like to thank my fellow researchers who gave constructive feedback and my family members who supported me at every step of my research.

Editor

A Review of the 1170 Andújar (Jaén, South Spain) Earthquake, Including the First Likely Archeological Evidence

J.A. Peláez, J.C. Castillo, F. Gómez Cabeza,
M. Sánchez Gómez, J.M. Martínez Solares and
C. López Casado

Additional information is available at the end of the chapter

1. Introduction

The origin of the town of Andújar (figures 1 and 2), in southern Spain, is likely a Roman settlement, as suggested by certain archaeological evidence in its historical center (figure 2). Andújar was probably founded to control a significant strategic route on the edge of the Guadalquivir River and ending in Córdoba. The town was a flat settlement, without natural shelters, presumably defended in this epoch by a defensive wall or fortification, although no evidence remains of it.

The first clear reference to the defensive wall of Andújar is a request from the emir '*Abd Allah* to the governor of the region, in the year 888, asking for aid to fortify the fort of *Anduyar* (Andújar) to protect the population from insurgents opposing the government of the Umayyad dynasty [1]. Subsequent archeological evidences indicate this fortification was restructured and extended in different epochs.

Based on information gathered from various archaeological digs in the historical center of Andújar (figure 2), we propose the following construction stages in its walled compound.

a) The emiral-caliphal town walls (the word emiral comes from Emirate, and caliphal from Caliphate, the two political systems existing during this epoch). They were built in the 9th and 10th centuries, when the presumable early fortification was established and extended to protect the population of the region. After that time, Andújar became one of the main towns in the countryside of Jaén, in the Guadalquivir Basin, significantly increasing in population,

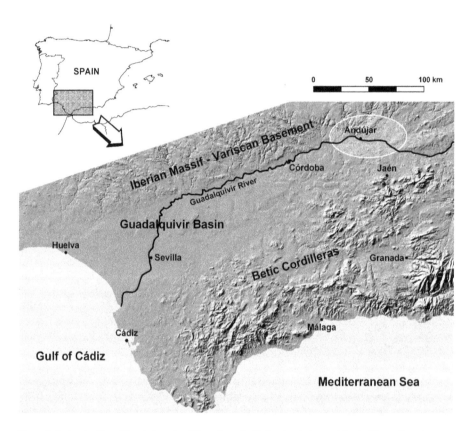

Figure 1. Regional setting of the study region. Ellipse shows the likely epicentral area of the Andújar earthquake.

and acquiring the condition of *Iqlim* (administrative district). The wall system is documented in an excavation carried out in the north of the town (figure 2) and revealing the ruins of a trapezoidal turret and several mud-walls of mortar with solid towers [2,3]. This mortar is extremely compacted gravel and lime, primarily white.

b) The taifa-Almoravid ramparts (taifas were small kingdoms in *Al-Andalus*, and the Almoravids were a Berber dynasty invader of Iberia, like the Almohads). During the 11th and the first half of the 12th centuries, the previous wall system was rearranged to incorporate the suburbs (recycling part of the obsolete emiral-caliphal walls) and thereby improving the security of the defensive system. These extensive works are an indication of the strategic nature of Andújar. Some ruins of the walls from this period have been documented in an archeological survey in the south of the town (figure 2). It includes a set of fortifications around one of the main gates of the city wall, the so-called *Puerta del Alcázar* (Alcazar gateway), the gateway for those entering town from the *Puente Romano* (Roman bridge). In this stage, the wall system was

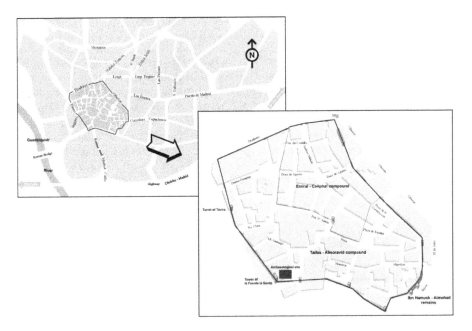

Figure 2. Current Andújar map. Historical center is enhanced showing the approximate extent of the different stages of ramparts and location of the archeological dig.

formed by mud-walls composed of small stones and lime in the externalmost part, and a mixture of materials inside, primarily dirt [4].

c) The *Ibn Hamusk*-Almohad stage (*Ibn Hamusk* was an insurgent who governed the region in agreement with Almoravids and Almohads, according to his interests). This was the most significant rearrangement of the wall system, carried out in the second half of the 12th and the beginning of the 13th centuries. Most remains discovered in the town are from this epoch [2,3]. Built structures at this stage are very homogeneous, constituting strongly tamped mud-wall made with lime, small stones, and sand. It is similar to those used in other fortifications in the northern Guadalquivir Basin (*e.g.*, Baños de la Encina, Giribaile and Santa Eufemia castles). In the northern sector of this wall system, in the most strategic spot, an alcazar (citadel) was built, which lasted until the beginning of the 20th century [5]. We are confident that this extensive rebuilding arose due to the deplorable conservation of the defensive wall system after the town was besieged and attacked various times in the second half of the 12th century and specially as a result of damage after the 1170 earthquake [6].

d) Christian stage or consolidation phase. Andújar was reconquered in 1225 by *Ferdinand III of Castile* and until the 15th century the ramparts were lightly repaired numerous times. Works mainly focused on cladding the most important structures of the defensive system, basically the towers and gateways, with sandstone ashlar. There are scarce elements remaining from this epoch, with the *Turret of Tavira* and the *Tower of 'de la Fuente la Sorda'* being worthy of note (figure 2) [5].

As mentioned above, after the taifa-Almoravid improvement and enlargement of the defensive wall system, a moderate earthquake struck this region. In the most recent Spanish earthquake catalog [7], it is termed the 1169, Andújar (Jaén) earthquake, the most destructive shock known in the whole Iberian Peninsula until the 1396, Tavernes de Valldigna (Valencia) earthquake (VIII-IX, macroseismic M_W 6.7) [8]. In the international scientific literature, the earthquake is known as the 1170, Córdoba earthquake [9]. This is due to the fact that these authors used only one of the contemporary manuscripts to catalog the event (ms. number 1 in Appendix), where Córdoba city is cited because it was much more important than Andújar and only 65 km away. In fact, Córdoba was the capital of the Emirate (750-929) and Caliphate (929-1031) of Córdoba, seat of emirs and caliphs. In the caliphate epoch, Córdoba reached a population of about 400000 inhabitants [10].

Although there is scarce information about the true effects of the earthquake, as we will see below, there is no doubt that there was a heavily damaging to destructive earthquake in Andújar. Moreover, in this paper we present what we consider to be the first archeological proof of this earthquake after interpreting the results obtained at an archeological site in the town. In archeoseismological studies, scientists must work bearing in mind the old rewrote saying: the presence of evidence is not evidence of presence. Considering contemporary documents of the event together with what we presume to be a recent archeoseismological result, we argue that in this case archeology supports the occurrence of this event.

This historical earthquake should be taken into account for future seismic hazard assessments in this region. If there is a moderate earthquake in an area, then there is a geological structure, known or unknown, that hosted it. Therefore, it is capable of being triggered again in future.

2. Contemporary written sources — Estimating effects and size

Three Arabic documents are the contemporary documentary evidence reporting the effects of this earthquake. They are two original manuscripts and a clear plagiarized summary of one of them providing no further information.

The best-known manuscript is that written by *Abû l-Walîd Muhammad Ibn Ahmad*, known during his lifetime as *Ibn Rushd* (the grandson), and later known in European literature as *Averroes*, a very important Andalusian philosopher and physician, among other activities. The text (ms. number 1 in Appendix) is included in his work *De Meteoris*, where the author revises different Aristotelian concepts. It must be taken into account that this text is not in fact a historic description of the incident, as in the other two manuscripts.

This is the only text considered in the historical seismicity works including this shock in references [9,11,12]. In reference [9], authors place the epicenter at Córdoba, assigning it a felt intensity (MM scale) equal to X. In contrast, in reference [12] authors place the epicenter at Andújar, assigning it a felt intensity (MCS scale) equal to IX.

It is important to note, as he states in his manuscript, that *Averroes* did not feel the mainshock himself, but several aftershocks when he went back to Córdoba from Sevilla. Concerning effects, it appears clear after reading the manuscript that: *a)* there was destruction and deaths, and *b)* there were aftershocks for three years. In addition, some statements suggest that the meisoseismal area, and presumably the epicentral area, is east of Córdoba, in the region of Andújar: *a)* effects were stronger east of Córdoba, and *b)* there were reported seismogeological effects in Andújar (*i.e.*, ground failure and/or ground liquefaction).

The second Arabic manuscript referring to this earthquake was written by *'Abd al-Malik b. Muhammad b. Ibn Sāhib al-Salā* (ms. number 2 in Appendix), an Arab chronicler contemporary of the event, and known by his translators for his explicitness, precision and fluency with respect to historical descriptions [13]. It is not totally clear but, it seems that *Ibn Sāhib al-Salā* was accompanying the caliph *Yūsuf I* and his brother the prince *Abū Sa'īd* on a journey throughout *Al Andalus*; accordingly, he probably felt the main quake or the aftershocks. In his manuscript, referring to events in the year 565 of the Arab calendar (September 25th, 1169 to September 13th, 1170), *Ibn Sāhib al-Salā* dramatically relates notable effects in Andújar: *a)* significant destruction, and *b)* a mainshock felt over a large region (the whole of *Al-Andalus*, and specifically in Córdoba, Granada and Sevilla).

The manuscript by *Ibn Sāhib al-Salā* has been selected to assign a date for this earthquake. He provides a detailed description for each year of the most important historic events in *Al-Andalus*. It is precisely this systematization in the timing of the related events that leads us to consider this date (January-February, 1170), as previously considered by the author in reference [14].

There are only two known works written by *Ibn Sāhib al-Salā* and only the second one, specifically the second volume of three of the second opus, called *Almohad caliphate history*, is currently preserved nowadays (in the Bodleian Library). This volume, which includes the manuscript referring to the Andújar event, covers 1159 to 1173. The only things known concerning the life of *Ibn Sāhib al-Salā* is what he relates about himself in this manuscript.

Further support as to the importance of this text and its chronicler, as stated in reference [13], is the fact that another contemporary historian, *Ibn 'Idārī*, copied and extracted literally (ms. number 3 in Appendix) the *Ibn Sāhib al-Salā* writings. Although it must be taken into account that the information in a derived source cannot always be considered confirmed information, in this case we agree with the author's [13] criterion.

Apart from these texts, there is no evidence of more reliable chronicles related to the event. A very short text (ms. number 4 in the Appendix) included in the so-called *Anales Toledanos* (Annals of Toledo), transcribed and compiled in reference [15], is the only other one worthy of note. They are contemporary medieval chronicles, characterized by extreme brevity and conciseness, reporting the most important occurrences of the time.

Some researchers in the Spanish historic seismicity consider that this citation likely refers to the Andújar earthquake [7,16], inferring that the quoted date is just the date of the event. In fact, as mentioned, in the recent Spanish earthquake catalog it is identified as the 1169, Andújar earthquake. Although it is quite possible that the Andújar earthquake was felt in the center of the Iberian Peninsula, we cannot guarantee that this sole reference in the *Anales Toledanos* can be taken as definitive proof for considering them to be the same event.

Evidently, the scarce documentary sources of this earthquake are a real problem in accurately dating the event. This lack also prevents researches from estimating with greater detail effects on buildings and people, establishing the meisoseismal area, and determining its impact on the society.

In a recent and comprehensive work [7], used as a basic historic seismic catalog in seismicity and seismic hazard studies in Spain, it is catalogued with a maximum intensity equal to VIII-IX (EMS-98 scale, used henceforth; [17]). As mentioned, it appears that there was ground liquefaction, implying at least a degree of intensity VIII. Nonetheless, this value must be supported independently of other effects. The intensity IX is sustained by the presumed effects on buildings, the result of considering that many houses were destroyed or collapsed and that many people died. Using the EMS-98 scale, this implies that many buildings of vulnerability class A (masonry structures of rubble stone, fieldstone, or adobe) sustained damage of grade 5 (total or near-total collapse). Quoted effects concerning mosque minarets described by *Ibn Sāhib al-Salā*, or damages in the ramparts, described below, as is well known, are very difficult to use for intensity assignment due to their complex structure and irregular behavior during an earthquake.

Using the empirical relationship among intensity and surface magnitude for the Mediterranean area in [18] gives M_s 6.0 ± 0.6 for this shock. In this estimate, evidently, possible site effects are not included.

An unpublished geophysical exploration test recently carried out by the authors in an alluvial terrace at the same level as Andújar using the H/V spectral ratio approach based on ambient vibrations, showed resonance frequencies in the range of 5-8 Hz, clearly related to very shallow structures, specifically a shallow sandy sedimentary layer. The potential amplification of earthquake motion by sediments in this area, using this or other approaches, must be explored in depth by future projects. Potential site effects in Andújar are expectable, increasing the seismic hazard in this location, but presumably decreasing the afore-mentioned expectable magnitude for the Andújar earthquake.

There are four fluvial terraces and the present flood plain in the Andújar area, with elevations above the river channel on the order of 55, 25, 13, and 6 m, from oldest to youngest [19,20]. These terraces comprise alluvial sediments from the Guadalquivir River with thicknesses ranging from approximately 5 to 10 m. They show a conglomeratic lithology with a silty mud-matrix that becomes sandy mud-matrix at the top of each terrace level. In general, the lower part of the terraces portrays the channel infilling and bar bedform, and the upper part shows the alluvial flood plain. The ages of these terraces range from the Holocene to 600 ka.

3. Seismic and geological framework

The only shock that stands out in the area is the studied earthquake. Seismicity is very scarce near Andújar, which is characteristic of the northern Guadalquivir Basin. Even within the basin, only a few minor earthquakes ($4.0 \leq M \leq 5.0$) are located. This region is therefore considered to have a low seismic hazard [21]. In fact, the design acceleration (called basic acceleration) for a return period of 500 years in the Spanish building code [22] for Andújar is only 0.05g.

The work in reference [21], using the spatially smoothed seismicity approach, includes a model with the most significant earthquakes in the Iberian Peninsula over the last 300 years. On the other hand, the design acceleration considered in the Spanish building code was computed through a typical zonified method using a broad seismic zone including the whole Guadalquivir Basin. Neither of these two assessments properly included the 1170 Andújar earthquake.

The only instrumental shock in this region deserving of mention is the March 10th, 1951 Linares earthquake (M_D 4.8, VIII), 25 km NE of Andújar [23]. In a recent work [24] this event is reevaluated (M_S 5.4, VI-VII, $h = 30$ km) and relocated to 20 km ESE of Andújar. These authors host this event at the base of the crust, in some deep fault near the southern boundary of the Paleozoic Iberian Massif related to the bending of the Paleozoic basement under the Neogene Guadalquivir Basin. More recently, in [25] has been reevaluated this earthquake once again (M_W 5.2, $h = 20$ km), relocating it to 70 km SSW of Andújar (*i.e.*, distancing it from Andújar). In this last work, the authors cannot explain the known macroseismic intensity distribution for this event after the epicenter relocation. Neither is a tectonic origin proposed.

The presumed mesoseismal area of the Andújar earthquake (figures 1 and 3) involves three tectonic and geographic domains (figure 3). The first, just north of Andújar, is the Paleozoic Iberian Massif, structured during the Variscan orogeny. The Iberian Massif has a flat topography, with the exception of its southern edge, which has a smooth slope.

The second domain is the Guadalquivir Basin, which is a classic foreland basin [26] formed by the collision of the Internal Zones of the Betic Cordillera with the southern paleomargin of the Variscan Iberian Massif. The Andújar area is located on the north side of the basin, at the piedmont of Sierra Morena. This range, along the Variscan border, has recently been interpreted as a flexural fore-bulge formed by the overload of the Betic chains above the Iberian crust [27].

The third tectonic domain is the frontal thrust belt (Subbetic and Prebetic) of the Betic Cordillera, which delineates an evident mountain front 40 km south of Andújar, a result of the aforesaid continental collision.

The most active stage of the continental collision occurred in the region between 20 and 7 Ma ago (Burdigalian-Tortonian) [28]. Nevertheless, there is clear evidence of recent tectonic activity along the Betic Mountain Front [29,30] that may account for its limited seismicity. This tectonic and regional seismic activity is probably related to the ongoing Africa-Iberia collision, with a convergence rate of 4-5 mm/year [31]. However, the Andújar region and the Betic

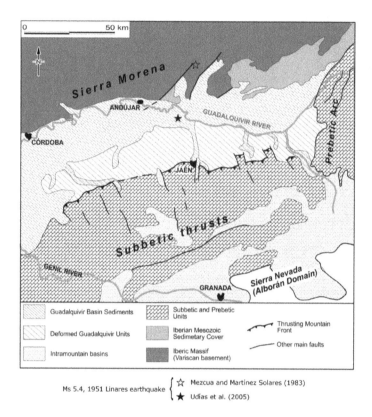

Figure 3. Tectonic sketch of the study region.

Mountain Front are relatively far from the current plate boundary. On the other hand, the Betic thrusts in the area are relatively shallow and detached from the Variscan Basement through the plastic Triassic materials (Keuper). Keuper sediments are rich in clay and gypsum, displaying very plastic behavior and lacking enough strength to accumulate a large amount of stress. Nevertheless, active Betic thrusts can account for small shallow earthquakes along the Betic Mountain Front [30].

Other tectonic structures near Andújar capable of accumulating enough stress to trigger moderate earthquakes, or at least displaying geomorphologic evidence of recent tectonic activity must also be considered. One possible source of the 1170 earthquake is the flexure of the entire lithosphere. It can cause moderate to strong earthquakes [32], but only from the beginning of the orogenic overload (Lower Miocene) until viscoelastic stress relaxation and equilibrium was reached, a few million years ago [33]. However, the present intraplate compression could lead to the amplification of the initial flexural foreland loading [34,35] and consequently the reactivation of seismic faults.

Another plausible seismic origin are the faults that fragmented the south Iberian crust during the Mesozoic, creating several blocks that produced swells and troughs in the marine paleo-margin [36]. These faults and their lateral ramps were tectonically inverted during the build-up of the Betic Cordillera (Miocene compression) and reutilized mainly as thrusts [37] until nowadays [38]. Thus, they continue to comprise crustal weak zones locally focusing the present crustal stress to host moderate earthquakes [30]. Most of these faults, seemingly with low slip rates, are now covered by the Guadalquivir Basin sediments (figures 1 and 3) and are difficult to recognize, even by geophysical exploration methods [39].

No Quaternary active faults have been described until now in the region and no clear limits can be traced at the lithospheric scale that could cluster the stress in the area. Nonetheless, any of the faults of these systems could explain a M_s 6.0 earthquake such as the Andújar event. Further work should look for active faults bordering rigid crustal blocks of the Variscan basement accommodating some of the present convergence between the Africa and Iberia plates.

4. Archeological evidence

A recent unfinished archeological survey in the south of Andújar, as previously mentioned, has revealed the ruins of a fortification that underwent rebuilding (figure 4). We presume that these repairs are related to the 1170 earthquake.

Figure 4. Current general view of the archeological site.

This archeological dig proves the existence of an early alcazar built in the 11th century, in the taifa-Almoravid stage, used approximately until the first half of the 12th century. In this epoch, it was replaced by a new alcazar located in the northern part of the town, built in the second half of the 12th and first half of the 13th centuries. After this, the early and obsolete alcazar was used only as a guard gate and control point of one of the main gateways (figure 5).

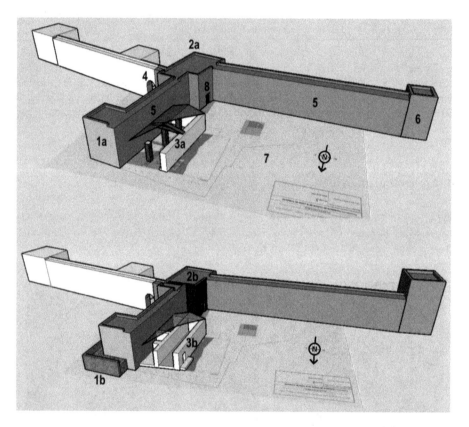

Figure 5. Reconstruction of the taifa-Almoravid (top) and *Ibn Hamusk*-Almohad (bottom) remains in the archeological site. 1a. Northern square solid tower. 1b. Reinforcement at the foundation of the tower. 2a. Southern complex tower. 2b. Reconstructed northwestern corner of tower. 3a. Rectangular building with adobe pillars. 3b. Reconstructed building, now including a dividing wall. 4. Gateway (Alcazar gateway). 5. Ramparts. 6. Western tower. 7. Courtyard. 8. Tower gate.

The defensive walls and towers of the early alcazar were built using a matrix of lime, sand, and small stones outside, and dirt and rubbish inside. This construction technique is unquestionably quick and cheap. However, although fillings of dirt and rubbish involve lower cost they also entail structural weakness, particularly for ground shaking during earthquakes. Two towers were excavated during the archeological survey, a small solid square tower to the north and another more complex one to the south (figure 5). This second tower had a room inside with a flat roof connecting the wall walks.

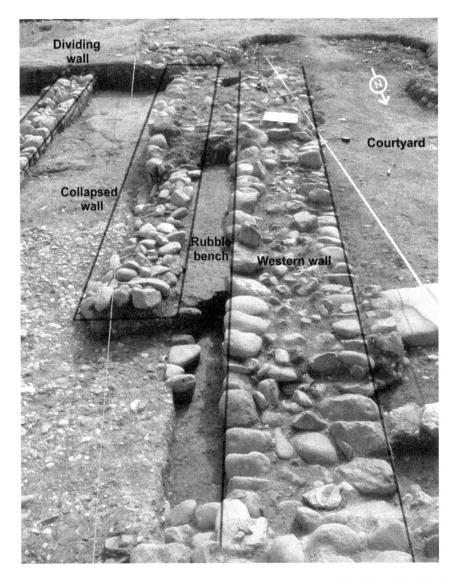

Figure 6. Details of the rectangular building within walls (3a and 3b in figure 5). Ordered fallen blocks of the collapsed wall are visible over the rubble bench.

Inside the alcazar, there was a large courtyard or ward where a simple rectangular building was probably used as a storehouse and/or kitchen (figures 5 and 6). The fact that it is not decorated suggests that it was not a room belonging to a palace or a residence. This building, attached to the eastern rampart, is 6.15 m wide, with an ashlar wall 1 m wide parallel to the rampart. The total length of this building is not completely known at the moment because the ends have not yet been excavated. Inside the building were found adobe pillars in a central position, probably related to the inward division and the roof support. Also, a rubble bench attached to the ashlar wall. The building was likely a space without interior walls divided into two rooms separated only by central pillars. These pillars held up central beams forming part of a roof of wood and tiles.

The 1170 earthquake affected (figure 5) both the fortification and the attached building, as we show below.

The northern tower was heavily damaged. In fact, later reinforcement of its foundation can be observed (figure 7). The reinforcement was made by means of a wall of tamped dirt 0.4 m wide and 1.1 m high surrounding the tower at a distance of 1.6 m. The space between the tower and the wall was filled in with cobblestones, two layers of dirt, and another of adobe. At present, the reinforcement can be clearly seen only in the southern part of the tower, but it apparently bordered the entire tower. Thus, the final surface of the tower was quite extended.

Figure 7. Detail of the uncovered reinforcement at the foundation of the northern tower (1b in figure 5).

The southern tower (figures 8 and 9) was also damaged. Specifically, its northwest corner was destroyed, likely the weakest part due to the opening of the gate. It was repaired, replacing the mud-wall by a tamped dirt wall, remodelling and decreasing the room inside it. The remodelled tower shows support with medium-sized rocks, barely preserved.

Figure 8. Details of the southern complex tower (2a in figure 5) showing the reconstructed corner. View from the courtyard.

The rectangular building inside the alcazar underwent near-total collapse (figure 6). The archeological dig found that the unattached western wall (1 m wide) of this building toppled inwards, to the east, and the fallen blocks are aligned the length of the wall. Fallen rocks tumbled on the rubble bench attached to the wall and on the floor. After the collapse, instead of cleaning out the blocks to the original floor, only a shallow cleaning was made. Therefore, collapsed blocks were buried, raising the level for a new floor. Moreover, previous pillars, likely very damaged or collapsed, were replaced by a dividing wall parallel to the fortification and to the front. From this dividing wall, the room was partitioned into four compartments after its reconstruction.

Figure 9. Details of the southern complex tower (2a in figure 5), showing the reconstructed corner. Opposite view of figure 8. 1. Early taifa-Almoravid wall and tower. 2. Reconstructed Ibn Hamusk-Almohad tower. 3. Christian period.

Some authors have used ordered fallen blocks as seismic-related kinematic indicators [40,41], among others effects, in order to determine, for example, the direction of seismic wave propagation or the degree of seismic shaking. In our case, the occurrence of just one episode unfortunately does not allow any conclusions to be inferred.

Since that time, this defensive complex likely had other functions, mainly as dwellings with rooms, kitchens, stables, and so on, as inferred from the archeological material found. This new use is supported by the fact that in the Almohad epoch, as noted previously, the alcazar was relocated to the northern part of the town, in a more strategic site [4].

Until now, these damages, reconstructions, and reinforcements could not be accurately dated. In any case, the fact that they occurred during the The *Ibn Hamusk*-Almohad stage (the second half of the 12th and the beginning of the 13th centuries), together with the documented date of the Andújar earthquake and the fact that no other likely historical explanation exists, supports a link between damages and the shock.

5. Summary and conclusions

In this paper we have presented a case study in seismic archeology that we believe to be the first likely archeological evidence of the 1170 Andújar earthquake. This case concerns one of the thorniest aspects of archeoseismology: to ascribe to historical attested earthquakes observed damages or effects in archaeological digs.

For this shock in southern Spain, only historical/documentary records have been available until now. Initially, a review of the scarce contemporary manuscripts was done, estimating some effects and justifying the presumed size. Then, damaged archeological structures and different repairs and reinforcements revealed in an archeological survey are proposed as true earthquake-related damages. In this case, in addition to the observed reinforcements and damages, there is the supporting evidence [42] of the historical record. We are confident that repairs and reinforcements in the two discovered and excavated towers, as well as the remodelling of a building attached to the rampart, including tumbled blocks along the length of its wall, are archeoseismological evidence. But it is still not possible from these effects to derive a better earthquake intensity estimate than that from contemporary manuscripts. In any case, we expect further results in future surveys, trusting that the site preserves additional traces of seismic activity in the ground.

The question still remains as to which geological structure hosted this shock. As discussed above, with no more plausible candidates, we suggest hidden faults bordering blocks of the basement as the most likely hypothesis.

Evidently, additional historical, archeological, and geological studies must be undertaken to estimate the size, effects and future implications of this earthquake.

Appendix

Manuscript number 1

author: *Abû l-Walîd Muhammad Ibn Ahmad, Ibn Rushd* (the grandson) or *Averroes*

source: *Taljīs kutub Aristātālīs fil-Hikma*. Cairo National Library

transcription: reference [43]

used translation: references [12,44]

Anyone who saw with his own eyes the earthquake which occurred at Córdoba in the year 566 [September 14th, 1170 - September 3th, 1171] has received confirmation [of the Aristotelian theory of earthquakes]. I was not at Córdoba at the time, and when I arrived, I heard the rumble which preceded the earthquake; people thought the rumble came from the west. I saw the earthquake being generated by the progressive movement of west winds. These earthquakes persisted at Córdoba throughout the year, and only ceased after about three years. The first earthquake caused great destruction and killed many people; it was said that at a place near Córdoba called Andujira [Andújar], the earthquake caused the earth to split open and something similar to ashes and sand came out of the fissure. To the east of Córdoba the effects were even more violent, whereas they were slighter to the west.

Manuscript number 2

author: *'Abd al-Malik b. Muhammad b. Ibn Sāhib al-Salā*

source: History of the Almohad Caliphate. Manuscript number 433. Bodleian Library. Oxford University

used transcription and translation: reference [13]

In the same year, the rain for the laid fields in al-Andalus was delayed until the Christian month of December, 1169, and [then] it rained and people sown. In this year big earthquakes happened at dawn and when noon declined in the month of *Ŷumadā al-ūlà* in the year that we chronicle [January 21th to February 19th, 1170], and they continued in the Andújar town for several days, until it almost dissapeared, and it was swallowed by the ground, and they continued, after this, in the Córdoba, Granada and Sevilla cities, and all *Al-Andalus*, and the eyewitness saw that walls of the houses shaken and sloped towards the ground, then they straighten and return to its position by the goodness of *Allah*, and because of that a lot of houses were destroyed in the quoted regions, and the minarets of the mosques.

Manuscript number 3

author: *Ibn 'Idārī*

source: *Al-Bayān al-Mugrib*. Manuscript discovered in a Koranic school in Tamagrūt, near Zagora, in the Draa Valley (Morocco)

used transcription and translation: reference [45]

In this year, a big earthquake happened, at dawn and at the end of the month of *Ŷumadā al-ūlà*, in part of *Al-Andalus*; the eyewitness sawn that walls were shaken and sloped towards the ground, but then they straighten and return to its position by the goodness of *Allah*. A lot of houses and minarets were destroyed in the Córdoba, Granada and Sevilla cities.

Manuscript number 4

source: vanished codex

used transcription: reference [15]

Toledo was shaken on February XVIII, MCCVII [February 18th, 1169] [The quoted data in the text concern to the Spanish or Hispanic era, or Era of the Caesars, beginning in the year 38 B.C.].

Acknowledgements

This work was mainly supported by the Seismic Hazard and Microzonation Spanish research group. The authors are grateful to Emanuela Guidoboni for their constructive comments in an early version of this manuscript.

Author details

J.A. Peláez[1*], J.C. Castillo[2], F. Gómez Cabeza[3], M. Sánchez Gómez[4], J.M. Martínez Solares[5] and C. López Casado[6]

*Address all correspondence to: japelaez@ujaen.es

1 Dpt. of Physics, University of Jaén, Jaén, Spain

2 Dpt. of Historical Heritage, University of Jaén, Jaén, Spain

3 CAAI (Andalusian Center of Iberian Archaeology), Jaén, Spain

4 Dpt. of Geology, University of Jaén, Jaén, Spain

5 Section of Geophysics, IGN (National Geographical Institute), Madrid, Spain

6 Dpt. of Theoretical Physics, University of Granada, Granada, Spain

References

[1] Guraieb JE. Al-Muqtabis de Ibn Hayyan (translation). Cuadernos de Historia de España 1952; XVII 155-166 (*in Spanish*).

[2] Salvatierra V, Castillo JC, Pérez MC, Castillo JL. The urban development in al-Andalus: The case of Andújar (Jaén). Cuadernos de Madinat al Zahra' 1991; 2 85-107 (*in Spanish*).

[3] Choclán C, Castillo JC. Immediate archaeological excavation in a property at 3 San Francisco St. and 12 Juan Robledo St., Andújar. In: Anuario Arqueológico de Andalucía, 1989. Sevilla, Spain: 1991. Vol. III, p319-327 (*in Spanish*).

[4] Castillo JC. Immediate archaelogical excavation carried out in a property located between the Alcázar, Altozano Deán Pérez de Vargas and Parras streets, in the town of Andújar (Jaén). In: Anuario Arqueológico de Andalucía, 1989. Sevilla, Spain: 1991. Vol. III, p276-291 (*in Spanish*).

[5] Eslava J. Castles in Jaén. Jaén, Spain: University of Jaén; 1999 (*in Spanish*).

[6] Peláez JA, Castillo JC, Sánchez Gómez M, Martínez Solares JM, López Casado C. The 1170 Andújar, Jaén, earthquake. A critical review. In: proceedings of the Fifth Spanish-Portuguese Meeting on Geodesy and Geophysics, 30 January - 3 February 2006, Sevilla, Spain (*in Spanish*).

[7] Martínez Solares JM, Mezcua J. Seismic catalog of the Iberian Peninsula (880 B.C.-1900). Madrid: Instituto Geográfico Nacional; 2002 (*in Spanish*).

[8]   Mezcua J, Rueda J, García Blanco RM. Reevaluation of historic earthquakes in Spain. Seismological Research Letters 2004; 75 75-81.

[9]   Poirier JP, Taher MA (1980). Historical seismicity in the near and middle east, north Africa, and Spain from arabic documents (VIIth-XVIIIth century). Bulletin of the Seismological Society of America 1980; 70 2185-2201.

[10] Guidoboni E, Comastri A, Traina G. Catalogue of ancient earthquakes in the Mediterranean area up to the 10th century. Rome-Bologne: Istituto Nazionale di Geofisica; 1994.

[11] Taher MA. Corpus des textes arabes relatifs aux tremblements de terre et autres catastrophes naturelles, de la conquête arabe au XII H / XVIII JC. LLD Thesis. University Paris I; 1979.

[12] Guidoboni E, Comastri A. Catalogue of earthquakes and tsunamis in the Mediterranean area from the 11th to the 15th century. Bologne-Rome: Istituto Nazionale di Geofisica e Vulcanologia; 2005.

[13] Huici A (1969). Ibn Sāhib al-Salā: Al-Mann Bil-Imāma. Valencia: Medieval texts collection, number 24; 1969 (in Spanish).

[14] López Marinas JM. The 1169 Al-Andalus earthquake. In: Basic seismic data determination for hidraulic infrastructures. Madrid: Dirección General de Obras Hidráulicas - Ministerio de Obras Públicas y Urbanismo; 1986 (in Spanish).

[15] Flórez E (1767). Sacred Spain. Vol. XXIII. Madrid; 1767 (in Spanish).

[16] Galbis J. Seismic catalog in the area inside meridians 5ºE and 20ºW Greenwich and the parallels 45º and 25ºN. Volume I. Madrid: Instituto Geográfico, Catastral y Estadístico; 1932 (in Spanish).

[17] Grünthal G., editor. European macroseismic scale 1998. EMS-98. Luxemburgo: Centre Europèen de Géodynamique et de Séismologie; 1998.

[18] D'Amico V, Albarello D, Mantovani E. A distribution-free analysis of magnitude-intensity relationships: an application to the Mediterranean region. Physics and Chemistry of the Earth (A) 1999; 24 517-521.

[19] Carral MP, Martín Serrano A, Sansteban JI, Guerra A, Jiménez Ballesta R. Determinant factors in the soil chronosequence in the morphodynamic evolution of the middle Guadalquivir (Jaén). Revista de la Sociedad Geológica de España 1998; 11 111-126 (in Spanish).

[20] Calero J, Delgado R, Delgado G, Martín García JM. SEM-image analysis in the study of a soil chronosequence on fluvial terraces of the middle Guadalquivir (southern Spain). European Journal of Soil Science 2009; 60 465-480.

[21] Peláez JA, López Casado C. Seismic hazard estimate at the Iberian Peninsula. Pure and Applied Geophysics 2002; 159 2699-2713.

[22] NCSR-02. Code of earthquake-resistant building: General part and construction. B.O.E. 2002; 244 35898-35967 (*in Spanish*).

[23] Mezcua J, Martínez Solares JM. Seismicity of the Ibero-Moghrebian area. Madrid: Instituto Geográfico Nacional; 1983 (*in Spanish*).

[24] Udías A, Muñoz D, Buforn E, Sanz de Galdeano C, del Fresno C, Rodríguez I. Reevaluation of the earthquakes of 10 March and 19 May 1951 in southern Spain. Journal of Seismology 2005; 9 99-110.

[25] Batlló J, Stich D, Palombo B, Macia R, Morales J. The 1951 Mw 5.2 and Mw 5.3 Jaén, southern Spain, earthquake doublet revisited. Bulletin of the Seismological Society of America 2008; 98 1535-1545.

[26] Galindo Zaldívar J, Jabaloy A, González Lodeiro F, Aldaya F. Crustal structure of the central sector of the Betic Cordillera (SE Spain). Tectonics 1997; 16 18-37.

[27] García Castellanos D, Fernández M, and Tornè M. Modeling the evolution of the Guadalquivir foreland basin (southern Spain). Tectonics 2002; 21 1018.

[28] Sanz de Galdeano C, Vera JA. Stratigraphic record and paleogeographical context of the Neogene basins in the Betic Cordillera, Spain. Basin Research 1992; 4 155-181.

[29] Sánchez Gómez M, Torcal F. Recent tectonic activity on the south margin of the Guadalquivir basin, between Cabra y Quesada towns (provinces of Jaén and Córdoba, Spain). In: proceedings of the Workshop in honour of the First Centennial of the Cartuja Observatory. One hundred years of Seismology in Granada, 28-29 September 2009, Granada, Spain. 2009 (*in Spanish*).

[30] Sánchez Gómez M, Peláez JA, García Tortosa FJ, Torcal F, Soler Núñez PJ, Ureña M (2008). Geological, seismic and geomorphological approach to the tectonic activity in the Eastern Guadalquivir Basin. In: proceedings of the Sixth Spanish-Portuguese Meeting on Geodesy and Geophysics, 11-14 February 2008, Tomar, Portugal. 2008 (*in Spanish*).

[31] De Mets C, Gordon RG, Argus DF, Stein S. Effect of recent revisions to the geomagnetic reversal time-scale on estimates of current plate motions. Geophysical Research Letters 1994; 21 2191-2194.

[32] McGovern PJ. Flexural stresses beneath Hawaii: Implications for the October 15, 2006, earthquakes and magma ascent. Geophysical Research Letters 2007; 34 L23305.

[33] García Castellanos D. Interplay between lithospheric flexure and river transport in foreland basins. Basin Research 2002; 14 89-104.

[34] Cloetingh S, Burov E, Beekman F, Andeweg B, Andriessen PAM, García Castellanos D, de Vicente G, Vegas R. Lithospheric folding in Iberia. Tectonics 2002; 21 doi: 10.1029/2001TC901031.

[35] Cloetingh S, Ziegler PA, Beekman F, Andriessen PAM, Matenco L, Bada G, García Castellanos D, Hardebol N, Dezes P, Sokoutis D. Lithospheric memory, state of stress

and rheology: neotectonic controls on Europe's intraplate continental topography. Quaternary Science Reviews 2005; 24 241-304.

[36] Vera JA. Evolution of the South Iberian Continental Margin. Mémories Musseum National Histoire Naturelle 2001; 186 109-143.

[37] Azañón JM, Galindo Zaldíbar J, García Dueñas V, Jabaloy A. Alpine tectonics II: Betic Cordillera and Balearic Islands. In: Gibbons W, Moreno T (eds.) Geology of Spain. London: Geological Society; 2002. p401-416.

[38] García Tortosa FJ, Sanz de Galdeano C, Sánchez Gómez M, Alfaro P. Recent tectonics in the Betic thrust front. The Jimena and Bedmar deformations (Jaén province, Spain). Geogaceta 2007; 44 59-62 (*in Spanish*).

[39] Ruano P, Galindo Zaldívar J, Jabaloy A. Recent tectonic structures in a transect of the Central Betic Cordillera. Pure and Applied Geophysics 2004; 161 541-563.

[40] Omuraliev M, Korjenkov A, Mamyrov E. Location of earthquake epicenters by non-traditional seismologic data. In: proceedings of the III Seminar "Non-Traditional Methods of Heterogeneity Study of the Earth Crust", Moscow, Russia. 1993 (*in Russian*).

[41] Korjenkov AM, Mazor E. Seismogenic origin of the ancient Avdat ruins, Negev Desert, Israel. Natural Hazards 1999; 18 193-226.

[42] Marco S. Recognition of earthquake-related damage in archaeological sites: Examples from the Dead Sea fault zone. Tectonophysics 2008; 453 148-156.

[43] Allah SF, Razik SA, editors (1994). Al-jawâmi_'fî l-falsafa: Kitâb al-âthâr al-'ulwîya. Cairo; 1994.

[44] Puig J (1998). Averroes, judge, physician, and Andalusian philosopher. Sevilla: Consejería de Educación y Ciencia (Junta de Andalucía); 1998 (*in Spanish*).

[45] Huici A. Ibn 'Idari: Al-Bayān al-Mugrib. New Almoravid and Almohad fragments. Valencia: Medieval texts collection, number 8; 1963 (*in Spanish*)

Chapter 2

Scaling Properties of Aftershock Sequences in Algeria-Morocco Region

M. Hamdache, J.A. Peláez and A. Talbi

Additional information is available at the end of the chapter

1. Introduction

This chapter is dedicated to the analysis of some aftershock sequences occurred in Algeria-Morocco region, namely Al Hoceima earthquakes of May 26, 1994 (Mw6.0) and February 24, 2004 (Mw6.1) which occurred in northern Morocco, the October 10, 1980 El Asnam earthquake (Mw7.3), the May 21, 2003 Zemouri earthquake (Mw6.9) and March 26 2006 Laalam earthquake (Mw5.2) in northern Algeria.

Aftershock sequence is usually attributed to the strain energy not released by the mainshock. Statistical properties of aftershocks have been extensively studied for long time. Most of them dealt with the distribution of aftershock in time, space and magnitude domain. Several authors have noted the importnace of systematic investigation of aftershock sequences to earthquake prediction and a number of statistical models have been proposed to describe seismicity characters in time, space and magnitude (Utsu, 1961; Utsu et al., 1995; Guo and Ogata, 1997; Ogata, 1999). It is also well known that aftershock sequence offer a rich source of information about Earth's crust and can provide and understanding of the mechanism of earthquakes, because tens of thousands of earthquakes can occur over a short period in small area. The tectonic setting and the mode of faulting are factors others than fault surface properties that might control the behavior of the sequence (Kisslinger and Jones, 1991; Tahir et al., 2012). It is widely accepted that aftershock sequence presents unique opportunity to study the physics of earthquakes. In the same time importnat questions concern the fundamental origin of some widely applicable scaling laws, as the Gutenberg-Richter frequency-magnitude relationship, the modified Omori law or the Omori-Utsu law for aftershock decay and Bath's law for the difference between the magnitude of the largest aftershock and mainshock.

The frequency-magnitude distribution (Gutenberg and Richter, 1944), firstly examined, describes the relation between the frequency of occurence and magnitude of earthquake

$$\log_{10}N(m) = a - bm \qquad\qquad (1)$$

where, $N(m)$ is the cumulative number of events with magnitude larger than or equal to m a and b are constants. The parameter a showing the activity level of seismicity exhibits significant variation from region to region becauses it depends on the period of observations and area of investigation. The parameter b describes the size distribution of events and is related to tectonic characteristics of the region under investigation. It is shown that the estimated coefficient b varies mostly from 0.6 to 1.4 (Weimer and Katsumata, 1999). Many factors can cause pertur- bations of the b value. For example it is established that the least square procedure introduce systematic biais in the estimation of the b parameter (Marzocchi and Sandri, 2003; Sandri and Marzocchi, 2005). The b value of a region not only reflects the relative proportion of the number of large and small earthquakes in the region, but is also related to the stress condition over the region. The physical implication of the b value, however, is not as obvious, is still under investigation.

The temporal decay of aftershocks is important task because it contains information about seismogenic process and physical condition in the source region. It is well known that the occurrence rate of aftershock sequences in time is empirically well described by the Omori- Utsu law as proposed by Utsu (Utsu, 1969). The power law decay represented by the Omori- Utsu relation is an example of temporal self-similarity of the earthquake source process. The variability of the parameter p-value is related to the structural heterogeneity, stress and temperature in the crust (Mogi, 1962; Kisslinger and Jones, 1991). The importance to derive reliable p-value, is due to the fact that it contains information about the mechanism of stress relaxation and friction law in seismogenic zones (Mikumo and Miyatake, 1979), but this information cannot be derived without a precise characterization of the empirical relations that best fit the data. In this study we examine the temporal patterns of aftershock distribution of earthquakes occurred in Algeria-Morocco region. The parameters of Omori-Utsu law are estimated by the maximum likelihood method, assuming that the seismicity follow a non- stationary Poisson process. Two statistical models, a first stage Omori-Utsu model, model without secondary aftershock and a two stage Omori-Utsu model including the existence of secondary aftershock activity, are tested for goodness of fit to aftershock data. We adopt the Akaike Information Criterion denoted AIC, (Akaike, 1974) as a measure for selecting the best among competing models, using fixed data.

A model describing the temporal behavior of cumulative moment release in aftershock sequences is analyzed as alternative approach with respect to the Omori-Utsu law. In this model the static fatigue is assumed to be the principal explanation of the aftershock temporal behavior. Following Marcellini (1997), the different aftershock sequences are analyzed using this model. The most important result shows that globally the model explain and fit correctly the data.

Modified Bath's law is analyzed for each aftershock sequence, it allows us to derive some important properties related to the energy partionning. This approach is used to derive the energy dropped during the mainshock.

The fractal dimension D_2 deduced from the correlation integral is used to carry out the spatial analysis of the studied sequences. The inter-event distances analysis is performed for the different aftershock sequences and the density of probability of the inter-event distances are derived using non-parametric approach, especially using the kernel density estimation technics (Silverma, 1986).

2. Tectonic sketch and seismicity overview

In this section we give a large overview of the tectonic skech and the seismicity of the studied region (Pelaéz et al., 2012). The Morocco-Algeria region, namely the Maghrebian region (Fig. 1) occupies the NW part of the African (Nubia) Plate in what is referred to its continental crust. Its oceanic crust continues till the area of the Azores Islands. To the North it is immediately situated the Eurasian Plate, although between the Gibraltar Arc and the South of Italy an intermediate complex domain is intercalated. This domain is formed by some oceanic basins, as is the Algero-Provençal Basin and the Thyrrehnian Basin, and by a former region, presently disintegrated and now forming the Betic-Rifean Internal Zone, the Kabylias (in Algeria), the Peloritani Mountains (Sicily) and the Calabrian area in Italy (AlKaPeCa domain). This area underwent from the early Miocene a northwards subduction of Africa, then opening the small oceanic basins quoted, accompanied by the disintegration of the AlKaPeCa domain.

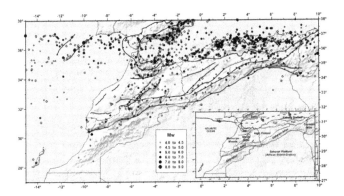

Figure 1. Tectonic sketch showing the main tectonic domains in the studied region.

Presently, the convergence between the Nubia Plate and Iberia has an approximate NNW-SSE direction, with values of the order of 3 to 5 cm/year, according the places. This compression is accompanied, at least in the area of the Gibraltar Arc (in the Alboran Sea) by a noticeable ENE-WSW tension, in some cases even more important than the compression. For this reason, in the Alboran area, some extensional movements can be important The Maghrebian region is also a complex area in which the Saharan Shield affected by the Pan-African Orogeny (Pre-cambrian to early Cambrian) is in contact with the Atlasic Mountains of mainly Alpine age

(Fig. 1). To the North of the Atlas is situated the Moroccan Meseta, to the West, and the High Plateaus in Algeria, which to the North contact with the Rif and Tell mountains, typically Alpine chains. The Saharan Shield forms part of the Precambrian areas of Africa, clearly cratonized and generally not affected by later important deformations. In fact, in the Maghrebian area it corresponds to a clearly stable area. In Morocco, the so called Antiatlas, corresponds to a Precambrian and, mainly, Paleozoic area, making a tectonic transition between the shield and the Atlas. The Atlasic Mountains correspond to an intracontinental chain. To the West, in Morocco, the High Atlas reach the coast in the Agadir area and continues to the Northeast and East, passing, although with lesser heights, to the Saharan Atlas, which cross Algeria and reach the central part of Tunis. They can be considered as aulacogens bordering the northern part of the Saharan shield. To the North, the Middle Atlas in Morocco has a different direction, NE-SW, separating the Moroccan Meseta and the High Plateaus, both forming by Paleozoic rocks, although with a Mesozoic and Tertiary cover, well developed in some areas. On the whole, the Atlasic Mountains has been tectonically unstable from the Triassic times, and along the Alpine orogeny suffered important deformations and, more recently, also important volcanism, reaching the Quaternary. The Rif and Tell thrust southwards the Moroccan Meseta and the High Plateaus, and even in some places part of the Atlasic Mountains. They are formed by sedimentary External zones (only slightly affected by metamorphism in some Moroccan places) and by Internal zones. Mostly Internal zones (divided in several tectonic complexes) are affected by alpine metamorphism, moreover the existence of previous Paleozoic and even older deformations. In any case, their present structure has being formed during the Alpine Orogeny. They appear mainly to the E of Tetuan, in Morocco, and in the Kabylias, in Algeria. These Alpine chains have being structured from the Cretaceous to the Oligocene-early Miocene. Later, were formed numerous Neogene basins, clearly cutting in many cases previous structures. In this time, particularly from the late Miocene to the present, a near N-S compression provoked the existence of strike-slip faults (NE-SW, sinistral, and NW-SE, dextral), moreover reverse fault, many of which has N70ºE to E-W direction. In many cases, the cited strike-slip faults moved mainly as normal faults, releasing by this way the regional tension, practically perpendicular to the compression.

In this study, we analyze the aftershock sequences triggered by some important and damaging earthquakes which occurred in the Morocco-Algeria region. The seismic activity in this region is mainly characterized by moderate to destructive magnitude events. It has been the site of numereous historical earthquakes, and was therefore subject to extensive damage and several thousands of casualties in the past. Among the largest recorded earthquake in Morocco, the February 29, 1960 Agadir (Morocco) earthquake (Ms6.0), the May 26, 1994 and February 24, 2004 Al Hoceima earthquakes with Mw6.0 and Mw6.1 respectively (Pelàez et al., 2007). In northern Algeria, earthquakes up to magnitude Ms 7.3 have been recorded, namely the October 10, 1980 El Asnam earthquake (Ms 7.3) (Ouyed et al.,1981) and May 21, 2003 Zemouri earthquake (Mw6.9) earthquake, (Hamdache et al., 2004). It is important to point out that these two events were the most damaging events occurred in northern Algeria and, the region of EL Asnam experienced in the past damaging earthquake on 09 September 1954 (Hamdache et al., 2010). The original epicentral database and the arrival time readings of the different aftershock sequences used, were obtained from the Spanish National Geographical Institut (IGN) earthquake catalog. For the sequences triggered by the May 21, 2003 Zemouri earthquake and by March 20, 2006 Laalam event, we used the data recorded by portable seismological stations network monitored by CRAAG. It is well known that the process to identify aftershock is related to the seismicity declustering process, a crucial step in separating an erthquake catalog into foreshock, aftershock and mainshock. This process is widely used in seismology, in particular for seismic hazard assessment and in earthquake prediction model. There are sevceral declustering algorithms that have been proposed. Up to now, most users have applied either the algorithm of Gardner and Knopoff (1974) or Reasenberg (1985). Gardner and Knopoff (1974) introduced a procedure for identifying aftershocks within seismicity catalog using interevent distances in time and space. They also provided specific space-time distances as a function of mainshock magnitude to identify aftershocks. This method is known as a window method and is one of the most used. Reasenberg's algorithm (1985) allows tolink up aftershock triggering within an earthquake cluster: if A is the mainshock of B, and B is the mainshock of C, then all A, B and C are considered to belong to one common cluster. When defining a cluster, only the largest earthquake is finaly kept to be the cluster's mainshock. In this study we have used the single-link cluster algorithm (Frohlich and Davis, 1990). We first identify large magnitude within the database. Earthquakes occurring within a certain distance d in km and period T in days after these potentialmainshock were extrated from the database. Aftershocks were then selected from this subset of events using single-link cluster algorithm (Frohlich and Davis, 1990). A space-time metric D is used to define the proximity of events relative to one another. The metric is defined as

$$D = \sqrt{d^2 + B^2 \Delta_t^2} \tag{2}$$

where d is the distance between two epicenters Δ_t is the temporal separation between two events and B is a constant relating time to distance. Afteshpocks are then defined as events within a D_c- size (critical size) cluster containing the mainshock. The single-link cluster analysis has been used with the parameters defined by Frohlich and Davis (1990), it is natural choice for aftershock selection because it avoids subjective decision-making with regard to the spatial distribution and duration of the sequence, it requires no assumption as the efficiency of event triggering, as is necessary when applying other clustering algorithms (e.g. Reasenberg, 1985).

The May 26, 1994 Al Hoceima earthquake (Mw 6.0) took place at 12 km depth, the focal mechanism indicates the presence of a main set of sinistral fault with a N-S trend, which may involve several parallel surface (Calvert et al., 1997). Analysis of the aftershock sequence highlighted the presence of NNE-SSW distribution of the seismicity (El Alami et al., 1998). The sequence used covers a period of about one year from the mainshock, including 318 events with magnitude ranged from 2.0 to 6.0. The February 24, 2004 Al Hoceima earthquake (Mw 6.1) took place at a depth of 10 to 14 km. The focal mechanism suggests that the active nodal plane corresponds to a sinistral strike-slip fault oriented N11 N and dipping 70 toward the E

(Stich et al., 2005). The aftershocks were aligned preferently in NNE-SSW, in the same way as one of the nodal planes of the focal mechanism. As pointed by Galindo-Zaldivar et al., (2009) the main stresses for the two aftershock sequences trigged by the May 26, 1994 and February 24, 2004 events have a trend with NW-SE compression (P-axis) and orthogonal NE-SW extension (T-axis) compatible with the present convergence of the Africa and Eurasia plate. The aftershock sequence includes 1233 events with magnitude ranged between 0.6 to 6.1. The time span of the sequence is about 1 year.

The October 10, 1980 El Asnam earthquake (Mw 7.3) is one of the most important and most damaging event occurred in northern Algeria. This event has taken place on the Oued-Fodda reversse fault. The later is segmented into three segments, ruptured along 26 km. This fault is located in the Cheliff high seismogenic Quaternary bassin, considered as very active. It is important to point out that almost all the seismicity in northern Algeria is located around the Plio-Quaternary intermountains active bassins (Meghraoui et al., 1996). The aftershock sequence used include the 130 most important magnitude events, ranged between 2.4 and 7.3. In the same way, the May 21, 2003 earthquake (Mw 6.9) has been located in the NE continuation of the south reversse fault system of the Quaternary Mitidja bassin (Maouche et al., 2011). The aftershock sequence we used include 1555 magnitude events, ranged between 0.9 to 6.9 and recorded during the first 40 days from the main shock (Hamdache et al., 2004). The March 20, 2006 Laalam earthquake (Mw 5.2), was located in the Babor chain, in the 'Petite Kabylie' south of Bejaia city. This chain belongs to the Tell Atlas, which is a portion of the Alpine belt in northern Africa. As pointed out by Beldjoudi et al., (2009) the region is affected by several faults. The regional seismicity analysis shows that the Babors chain seems to belongs to a "transition zone" between a large belt of reverse faulting along the western and central part of northern Algeria and a more distributed zone where deformation is mainly accomodated through strike-slip faulting (Beldjoudi et al., 2009). The aftershock sequence include 111 of the best recorded events with more than 54 with $RMS < 0.3s$, $ERH < 3km$ and $ERZ < 3km$. The event magnitude varies between 1.3 and 5.2.

3. Magnitude-frequency relationship

The frequency-size distribution (Gutenberg and Richter, 1944) describes the relation between the frequency of occurrence and the magnitude of earthquake

$$\log_{10}N(m) = a - bm \quad ; \quad m_c \leq m \tag{3}$$

the statistic $N(m)$ is the cumulative number of events with magnitude larger than or equal to m, whereas a and b are positive constant. The parameter a which exhibits significant variations in space measures the activity level of seismicity. This parameter is sensitive to the input period of observation and area of investigation. The parameters b describes the size distribution of events and is related to tectonic characteristics of the region under investigation. It is shown that the estimated coefficient b varies mostly from 0.6 to 1.4 (Weimer and Katsumata, 1999).

Many factors can bias the b value estimation. It is established for example that the least square procedure introduce systematic error in the estimation of the b parameters (Marzocchi and Sandri, 2003; Sandri and Marzocchi, 2005). Locally, the b value not only reflects the relative proportion of the number of large and small earthquakes in the region, but it is also related to the stress condition over the region (Mogi, 1962). The physical meaning of the b value is, however not clear and still under investigation.

The first step in the analysis of the Gutenberg-Richter law, is the determination of the threshold magnitude of completness. It is defined as the lowest magnitude for which all the events are reliably detected (Rydelek and Sacks, 1989). There are many approaches to the estimation of m_c, the most popular are;

The maximum curvature (MAXC) method (Weimer and Wyss, 2000) defines the completeness magnitude as the magnitude for which the first derivative of the frequency magnitude curve is maximum (being the maximum of the non-cumulative frequency-magnitude distribution). The goodness of fit (GFT) method (Weimer and Wyss, 2000) compares the observed frequency-magnitude distribution with a synthetic distribution, and the goodness of fit is calculated as the absolute difference of the number of earthquake in each magnitude bins between the observed and synthetic distribution. The synthetic distribution is calculated using a and b-values estimated from the observed dataset by increasing the cutoff magnitude. The completness magnitude, is given by the magnitude for which 90% of the data are fitted by a straight line. The entire magnitude range (EMR) method was developed by Ogata and Katsura (1993) and modified later by Woessner and Weimer (2005). The maximum likelihood estimation method is used to estimate the power law G-R law parameters a and b. The same method is applied to estimate the mean and standard deviation of the Normal distribution considered for the incomplete part of the frequency-magnitude distribution. μ and σ are the magnitude at which 50% of the earthquakes are detected and the standard deviation describing the width of the range where earthquakes are partially detected, respectively.

The frequency-magnitude distribution as shown previously is defined as,(Gutenberg and Richter, 1954)

$$Log_{10}N(\geq m) = a \quad - \quad bm \tag{4}$$

where, $N(\geq m)$ is the cumlulative number of earthquakes with magnitudes greater than or equal to magnitude m, and a and b are constants. This relationship holds for global earthquake catalogs and is applicable to aftershock sequences as well. The estimation of the b- value has been the subject of considerable research, and various methods exist. The most statistically appealing of these is the maximum likelihood method first used independently by Aki (1965) and Utsu (1965), which gives the estimate of the b- value as

$$\hat{b} = \frac{\log_{10}e}{m - m_{min}} \tag{5}$$

where m_{min} is the minimum magnitude up to which the Gutenberg-Richter law can accurately represent the cumuluative number of earthquakes larger than or equal to a given magnotude, and $\bar{m}$ is the average magnitude. The definition of m_{min} origionates from the definition of the probability density function for magnitudes, $f(m)$, consistent with the mathematical form of the Gutenberg-Richter law. This is given as (Bender, 1983)

$$f(m) = \frac{\lambda \exp\left[-\lambda\left(m - m_{min}\right)\right]}{1 - \exp\left[-\lambda\left(m_{max} - m_{min}\right)\right]} \tag{6}$$

where m_{max} is the maximum magnitude up to which $f(m)$ describes the distribution of magnitudes and $\lambda = bLn(10)$. There are two problems with equation (4) when applied to real earthquake catalogs. Firstly, it considers reported earthquake magnitudes as a continuous variable, which is not accurate as most earthquake catalogs report magnitudes up to a precision of one decimal place for each event. Therefore, magnitude should be considered as a grouped or binned variable with a finite non zero bin length δm, which generally and in our case it is taken equal to 0.1, $\delta m = 0.1$. This means that instead of observing m_{min} physically in the catalog, we observe a different minimum magnitude $m_c = m_{min} + \dfrac{\delta m}{2}$, which is the aforementioned completeness magnitude. This is the smallest observed magnitude in the catalog above which the cumulative number of earthquakes, above a given magnitude, are accurately described by the Gutenberg-Richter law. It is very important to note that the minimum magnitude of completeness m_c, physically observed in the catalog, is not present in the equation (4), and hence we need to express m_{min} in terms of m_c to be able to use the formula for real catalog. Assuming that $m_{min} = m_c$ would lead to a bias in the estimate. The other source of inaccuray in equation (4) is the fact that it considers the maximum magnitude value for the dataset to be infinite, which is never the case. in fact, for aftershock sequences, the maximum magnitude in the sequence can be pretty small. This too introduces a bias in the estimate of the b- value obtained using the equation (4) (Bender, 1983; Tinti and Mulargia, 1987; Guttorp and Hopkins, 1986). This problem was considered by Bender (1983), and she gave a method for obtaining the b- value of grouped and finite maximum magnitude data as a function of the root of the following equation

$$\frac{q}{1-q} + \frac{nq^n}{1-q^n} = \frac{\bar{m} - m_{min} - \dfrac{1}{2}\delta m}{\delta m} \tag{7}$$

where $q = \exp[-Ln(10).\hat{b}.\delta m]$ and $n = \left(m_{max} - m_{min}\right)\big/ \delta m$.

It is straightforward to see that for $n \rightarrow \infty$, equation (6) gives (Tinti et al.,1987; Gutorp et al., 1986).

$$\hat{b} = \frac{Log_{10}e}{\delta m} Ln\left(1 + \frac{\delta m}{m - m_c}\right) \qquad (8)$$

(Tinti and Mulargia, 1987; Gutorp and Hopkins, 1986). It has been observed, however, that for a difference of about 3.0 in magnitude between m_{min} and m_{max}, the value of b obtained from equation (6) agrees closely with the asymptotic value of $\hat{b}$ obtained from equation (7) (Bender, 1983).

For the two aftershock sequences triggered by the mainshocks occurred around Al Hoceima city in Morocco on 1994 and 2004, the threshold magnitude m_c has been examined in details, using the different procedures introduced previously, especially the maximum curvature procedure MAXC (Weimer and Wyss, 2000) and the changing point procedure introduced by Amores, (2007). The results obtained for these two aftershock sequences, are shown in Fig. 2

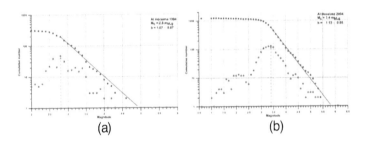

(a) (b)

Figure 2. Graphs showing the non cumulative number of events, the threshold magnitide and the adjustment of the cumulative number by straight line with equation $Log_{10}N(\geq m) = a - bm$ for $m \geq M_c$. (a) For Al Hoceima 1994 afeshock sequence and (b) for Al Hoceima 2004 aftershock sequence.

In Fig.2, the frequency-magnitude relation for the 1994 and 2004 aftershocks series of Al Hoceima (Morocco) are displayed. Based on maximum curvature procedure (MAXC), the magnitude of completeness was taken equal to 2.8 for the 1994 aftershock seqiuence and 3.4 for the 2004 sequence. It is important to point out that the changing point procedure (Amores, 2007) gives the same results. Using these threshold magnitudes, we derive the b-value of the Gutenberg-Richter relationship and its standard deviation using the maximum likelihood procedure. The b-value is estimated equal to 0.92 ± 0.02 for the 1994 aftershock sequence and 1.073 ± 0.003 for the 2004 series. For both sequences the b-value obtained is close to 1.0, the typical value for aftershock sequence. The obtained parameter a is equal to 4.689 ± 0.058 for the 1994 aftershock series and equal to 6.305 ± 0.014 for the 2004 aftershock sequence. The results obtained for the aftershock sequences triggered by the events occurred in northern

Algeria, namely the October 10, 1980 El Asnam earthquake (Mw 7.3), the May 21, 2003 Zemouri earthquake (Mw 6.9) and March 20, 2006 Laalam earthquake (Mw 5.2) are displayed in the following Fig. 3.

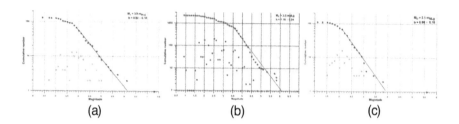

(a)   (b)   (c)

Figure 3. Graph displaying the non cumulative number of event in blue circle and the adjustment of the cumulative number (red circles) by straight line, representing the Gutenberg-Richter relation for the aftershock sequence triggered by (a) October 10, 1980 El Asnam earthquake (b) by May 21, 2003 Zemouri earthquake and (c) by March 20, 2006 Laalam earthquake.

For the sequence of October 10,1980 El Asnam earthquake and March 20, 2006 Laalam earthque, we have used the maximum curvature procedure to derive the threshold magnitude, we obtained $m_c = 3.9\, m_{bLg}$ for the El Asnam 1980 aftershock sequence and $m_c = 2.1\, m_{bLg}$ for Laalam 2006 sequence. The higher value obtained for El Asnam series reflect the quality of the data include in the El Asnam series, it seems that the file doesn't include all the events. It is till now one of the "best" file including the mainshock and almost all the aftershock occurred just after the mainshock. The b-value obtained for these two series, 0.82 ± 0.10 for El Asnam series and 0.99 ± 0.10 for Laalam aftershock series are close to 1.0, a typical value for the aftershock sequences. For the Zemouri aftershock sequence we have used the best combination between m_c obtained for 95% and 90% confidence and the maximum curvature procedure (Weimer and Wyss, 2000), which gives a threshold magnitude equal to $m_c = 3.5\, m_{bLg}$. To improve the b-value analysis for this sequence, we obtained a threshold magnitude $m_c = 3.7\, m_{bLg}$, using the procedure based on the stabilisation of the b-value and its uncertainty derived by Shi and Bolt (1982), as implemented by Weimer and Zuniga, (1994) in the software Zmap under Matlab. We obtained $b = 1.10 ± 0.04$ for the first procedure and $b = 1.30 ± 0.06$ for the second procedure, the difference between the two values is of the order of 0.20.

4. Decay rate of aftershock activity — Omori-Utsu law

The third studied scaling law is related to the decay rate of aftershock activity. It is well know that the decay rate of aftershock activity with time is governed by the modified Omori law or Omori-Utsu law (Utsu et al. 1995),

$$\lambda(t) = \frac{k}{(t+c)^p} \qquad (9)$$

$\lambda(t)$ is the rate of occurrence of aftershocks per unit time, at time t after the mainshock ($t = 0$). The parameters k, c and p are positive constants. k depends on the total number of events in the sequence, c on the rate of activity in the earliest part of the sequence and p is related to the power law decay of aftershocks. It is generally accepted that the number of aftershocks cannot be counted completely in the beginning of the sequence when small shock are often obscured by large ones due to overlaping, thus an overly large value of the parameter c is obtained. After Utsu, (1971), the parameter c must be zero if all shocks could be counted. Two opinions constitue the nowdays debate around the c value: one is that c value is essentially zero and all reported positive c value results from incompletness in the early stage of an aftershock sequence; the second point of view is that c value do exist (Enescu and Ito, 2002; Kagan, 2004; Shcherbakov et al. 2004a, b; 2005, 2006; Enescu et al. 2009). The constant c is a controversial quantity (Utsu et al. 1995; Enescu et al. 2009) and is mainly influenced by the incomplete detection of small aftershocks in the early stage of the sequence (Kisslinger and Jones, 1991). According to Olsson (1999), p values generally vary in the interval 0.5 to 1.8 and this index has shown by Utsu et al. (1995), differs from sequence to sequence and vary according to the tectonic condition of the region, however it is not clear which factor is the most significant in controlling p value. Although more attention in the estimation of the p value have been done to take into account the recommandation by Nyffenegger and Frolich (1998, 2000), we have use the standard use the standard deviation obtained by maximum likelihood which could be over or underestimating (Nyffenegger and Frolich, 1998; 2000).

In this study aftershock sequences are modeled using point process defined by the following conditional intensity,

$$\lambda(t|\mathfrak{J}_t) = \frac{k}{(t+c)^p} \qquad (10)$$

with, $\mathfrak{J}_t$ is the history of observation until time t, e.g the σ- algebra generated by the events occurred before the time t, thus, following Daley and Vere-Jones (2005), the family $(\mathfrak{J}_t, \ t \in \mathbb{R}_+)$ is called the natural filtration of the point process or the internal history. The conditional intensity in equation 4, is independent of the internal history and depends only on the current time t, like $\lambda(t \mid \mathfrak{J}_t)$. It defines a non-stationary (nonhomogeneous) Poisson process. We assume the occurrence time's of the aftershock sequence, namely $\{t_1, \ t_2, \ ...., \ t_n\} \subseteq [T_s, \ T_e]$ are distributed according to a non-stationary Poisson process with conditional intensity function defined by equation 4. The parameters k, p and c are estimated by maximizing the following likelihood function

$$L\left(\theta \big| T_s, T_e\right) = \left\{ \prod_{i=1}^{i=n} \lambda\left(t_i, \theta \big| \mathfrak{I}_{t_i}\right) \right\} \exp\left\{ -\int_{T_s}^{T_e} \lambda\left(t, \theta \big| \mathfrak{I}_t\right) dt \right\} \tag{11}$$

where, $\theta = (k, p, c)$ are the model parameters.

The log likelihood is then given by;

$$Ln\, L\left(\theta \big| T_s, T_e\right) = nLn(k) - p\sum_{i=1}^{i=n} Ln\left(t_i + c\right) - k\Psi\left(\theta \big| T_s, T_e\right) \tag{12}$$

with

$$\Psi\left(\theta \big| T_s, T_e\right) = \begin{cases} \dfrac{\left(T_e + c\right)^{1-p} - \left(T_s + c\right)^{1-p}}{1-p} & \text{for } p \neq 1 \\[4mm] Ln\left(T_e + c\right) - Ln\left(T_s + c\right) & \text{for } p = 1 \end{cases} \tag{13}$$

it follows that the maximum likelihood estimate MLE $\theta = (\tilde{k}, \tilde{p}, \tilde{c})$ of the parameter θ is solution of the following equation

$$\frac{\partial}{\partial \theta} Ln\, L\left(\theta \big| T_s, T_e\right) = 0 \tag{14}$$

Following, Ogata (1983), the maximum estimation of the parameters are obtained by using the Davidon-Fletcher-Powell optimization algorithm (Press et al. 1986, pages 277, 308) applied to equation 14. The standard deviation of the estimated parameters through the maximum likelihood procedure are derived using the inverse of the Fisher information matrix $J(\theta)^{-1}$. In our case, Fisher information matrix is given by

$$J(\theta) = \int_{T_s}^{T_e} \frac{1}{\lambda\left(t, \theta \big| \mathfrak{I}_t\right)} \frac{\partial}{\partial \theta'} \lambda\left(t, \theta' \big| \mathfrak{I}_t\right) \frac{\partial}{\partial \theta} \lambda\left(t, \theta \big| \mathfrak{I}_t\right) dt \tag{15}$$

the results obtained using this procedure for the aftershock sequences are shown on Fig. 4.

It is often observed that a sequence of aftershocks contains secondary aftershocks, which are aftershocks of a major aftershock (Utsu, 1970). Secondary aftershock are typically detected as changing-point in the activity rate of the sequence using Akaike information criteria (AIC) (Akaike, 1974). In this study, the changing-point in the activity rate of the sequence is detected by using the plot of cumulative number of events vs time from the mainshock. Assuming one secondary aftershock occurred at time τ_0, we test four different hypothesis, H1 : no secondary aftershock, H2: a secondary aftershock series does exist with the same p-value and c- value as the original series, H3; secondary aftershock series does exist with same c- value as the original series and H4: a secondary aftershock series does exist with Omori law parameters different from the original series. Four point process models are tested. Respectively, the conditional intensity and the cumulative function of such point process are given by;

$$\lambda(t|\Im_t) = \begin{cases} \dfrac{k_1}{\left(t+c_1\right)^{p_1}} & for \quad T_s \le t < \tau_0 \\[4mm] \dfrac{k_1}{\left(t+c_1\right)^{p_1}} + \dfrac{k_2}{\left(t-\tau_0+c_2\right)^{p_2}} & for \quad \tau_0 \le t \le T_e \end{cases} \tag{16}$$

and the cumulative function is given by;

$$N(t) = \begin{cases} \dfrac{k_1}{\left(1-p_1\right)}\left[\left(t+c_1\right)^{1-p_1} - c_1^{1-p_1}\right] & for \quad T_s \le t < \tau_0 \\[4mm] \dfrac{k_1}{\left(1-p_1\right)}\left[\left(t+c_1\right)^{1-p_1} - c_1^{1-p_1}\right] + \dfrac{k_2}{\left(1-p_2\right)}\left[\left(t-\tau_0+c_2\right)^{1-p_2} - c_2^{1-p_2}\right] & for \quad \tau_0 \le t \le T_e \end{cases} \tag{17}$$

$T_s = 0$ is the occurrence time of the main shock, and T_e is the occurrence time of the last event. All the occurrence time are counted as the number of days elapsed from the mainshock. For all the model relative to the hypothesis H1, H2, H3 and H4, the model with the lowest AIC is selected as the best model. The AIC is used as a measure to select which model fits the observations better. This is a measure of which model most frequently reproduces features similar to the given observations, and is defined by

$$AIC = (-2)\,Max(Ln\,likelihood) + 2(number\ of\ used\ parameters) \tag{18}$$

a model with smaller value of AIC is considered to be better fit to the observations. The fit of the data by the Omori-Utsu law is obtained for both the minimlum magnitude m_{min} in the

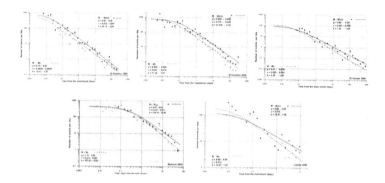

Figure 4. Fit of the number of events pêr day by the modified Omori law, for the studied aftershock sequence. The graphs display the fit for $m \geq m_c$ in dashed line and in solid line for $m \geq m_{min}$.

	Omori-Utsu parameters for $m \geq m_c$		
	$p \pm \sigma_p$	$c \pm \sigma_c$	$k \pm \sigma_k$
Al Hoceima 1994	0.76 ± 0.03	0.0040 ± 0.0008	12.47 ± 1.51
Al Hoceima 2004	0.85 ± 0.02	0.034 ± 0.016	47.38 ± 3.81
El Asnam 1980	0.84 ± 0.06	0.025 ± 0.045	4.48 ± 3.81
Zemouri 2003	1.13 ± 0.06	0.311 ± 0.053	107.53 ± 15.83
Laalam 2006	0.69 ± 0.09	0.014	10.27 ± 1.47

Table 1. Omori-Utsu Parameters p, c, k and their standard deviation obtained for each aftershock sequence with magnitude above the threshold magnitude, $m \geq m_c$

aftershock sequence and the threshold magnitude m_c discussed and derived in the previous section. The results of the fit are shown on the following figure

The Omori-Utsu parameters obtained for each aftershock sequence for magnitude above the threshold magnitude, $m \geq m_c$, are given on the Table 1.

As pointed previously, it is often observed that a sequence of aftershocks contains secondary aftershocks, aftershocks of a major aftershock (Utsu, 1970). If a secondary aftershock equence starts at time τ_0 then from the Eq. 15 and 16, the conditional intensity of the point process is given as

$$\lambda(t|\Im_t) = k_1\left(t + c_1\right)^{-p_1} + H\left(t - \tau_0\right)k_2\left(t - \tau_0 + c_2\right)^{-p_2} \tag{19}$$

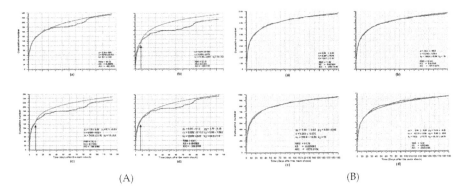

(A)   (B)

Figure 5. Graphs showing the cumulative number of aftershocks of Al Hoceima earthquake of 1994, with magnitude greater or equal to m_c, fitted by a simple Modified Omori law – graph (a)- and two stage Modified Omori law, including secondary aftershock shown on the plot – graphs b, c and d – The parameters of each model and the AIC are given on the plot. (A) for AL Hoceima 1994 aftershock sequence and (B) for AL Hoceima 2004 afetsrhock sequence.

denoting, $\theta = (k_1, \ p_1, \ c_1, \ k_2, \ p_2, \ c_2, \ \tau_0)$ and where $H(t)$ is teh Heaviside unit step function. The occurrence of secondary aftershock is analyzed using the procedure of detecting the aftershock activity change point by AIC, procedure introduced by Ogata (1999), Ogata (2001) and Ogata et al., (2003). Using the changing point AIC procedure to the aftershock sequence of Al Hoceima 1994, it appears that the time $t = 8.0215$ days is a suspecious point of activity change. Fitting the data by two stage Omori-Utsu models, shows that the smaller AIC value, equal to -403.6618 is obtained for simple Omori-Utsu model. The results obtained are shown on Fig. 5. The results obtained for the two stage Omori-Utsu models are shown on Fig 5 (b), (c) and (d). The AIC values are include on the graphs.

The analysis of the aftershock sequence of Al Hoceima 2004, give us that a suspecious point of aftershock activity change is given by $\hat{t} = 2.402$ days, the lower AIC equal to $AIC = -3787.1548$ is obtauined for the simple Omori-Utsu model as shown on the following Fig. 6(B). The results obtained for the two stage Omori-Utsu models are displayed on Fig. 5 (b), (c) and (d), including the AIC values. The principal reason of this, seen from the comparaison of the predicted and real cumulative curves would be that the secondary aftershock triggered by large aftershocks are not frequent enough in relation to their magnitude. It is well established that the residual analysis of point process is a "good" tool to evaluate the goodness of fit of the selected model to the data (Ogata, 1988; Ogata, 1999). The integral

$$\Lambda(t) = \int_0^t \lambda(s|\mathfrak{I}_s)ds \qquad (20)$$

of non-negative conditional intensity function produces a 1-1 transformation of the time from t to $\tau = \Lambda(t)$, so that the occurrence times $t_1, t_2, t_3, t_4, t_5, \ldots\ldots\ldots, t_N$ are transformed into

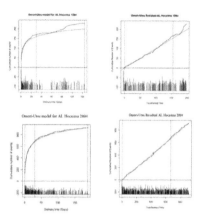

Figure 6. Residual analysis to derive the best fitting for the two Al Hoceima aftershocks.

	Akaike Information Criteria				
	Model 1	**Model 2**	**Model 3**	**Model 4**	**B. Model**
Al Hoceima 1994	-639.1429	-609.7559	-609.276	-633.7404	Model 1
AlHoceima 2004	-3616.8231	-36020.7106	-3609.2007	-3610.6351	Model 1
El Asnam 1980	-47.1127	-32.118	-31.2841	-42.9912	Model 1
Zemouri 2003	-8973.1703	-9244.9968	-9342.0748	-9353.1883	Model 4
Laalam 2006	-293.5383	-274.8197	-275.9812	-288.698	Model 1

Table 2. Akaike information criteria (AIC) of each model and for each aftershock sequence. Table gives in the last column the best model with the lower AIC.

$\tau_1, \tau_2, \tau_3, \tau_4, \tau_5, \ldots\ldots\ldots\ldots\ldots, \tau_N$. It is well known that τ_k, $k = 1, \ldots, N$ are distributed as a standard Poisson process. If $\lambda(t \mid \Im_t)$ is the intensity function of the process actually generating the data, using the maximum likelihood conditional intensity $\lambda_{\hat{\theta}}(t \mid \Im_t)$, the corresponding $\hat{\tau}_1, \hat{\tau}_2, \hat{\tau}_3, \hat{\tau}_4, \hat{\tau}_5, \ldots\ldots\ldots\ldots, \hat{\tau}_N$ called residual process (Ogata, 1988; 1999), provides a measure of teh deviation of the data from the hypothezed model. The residual analysis performed for the two aftershock sequences, trigged respectively by Al Hoceima earthquakes of 1994 and 2004 is shown on Fig. 6. The graphs display the Omori-Utsu model fit and the residual process for Al Hoceima 1994 series and for Al Hoceima 2004 series.

from the graphs of the residual process we can deduce, as shown previously using the *AIC* criteria, a global agreement with the Omori-Utsu model. We proceded with the same technics for the others aftershock sequences, the *AIC* is used to select the most appropriate model fitting the data. Table 2, gives the obtained results.

Following figure 7.6 displays for El Asnam 1980, Zemouri 2003 and Laalam 2006 aftershock sequences the adjustment of the data with the appropriate model, as deduce from the Table2

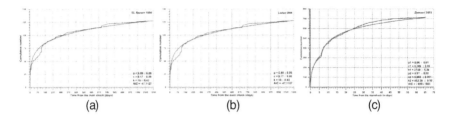

(a)   (b)   (c)

Figure 7. Adjustment of the cumulative number of events by the appropriate Omori-Utsu model, first stage Omori-Utsu model for (a) EL Asnam 1980 aftershock sequence and (b) for Laalam aftershock sequence. Two stage Omori-Utsu model for Zemouri 2003 aftershock sequence

The temporal aftershock decay is also analyzed using the approach introduced by Marcellini (1997). This approach describes the temporal behavior of the cumulative seismic moment released in aftershock sequences. It is an alternative approach to the Omori-Utsu law, previously analyzed. Static fatigue is assumed to be the principal explanation of the aftershock temporal behavior. Under the condition that the main shock causes a redistribution of stress, the initial stress condition of the afterhock sequence at main shock origin time t_0 can be considered as the superposition of the stress before the main shock and the stress step $d\sigma$ caused by the dynamic rupture of the main shock. It has been shown Marcellini (1995) that

$$S(t_i) = d\sigma \; + \; \frac{RT}{\gamma} Ln(t_i) \tag{21}$$

where t_i is the time from the mainshock to the *ith* aftreshok, T the absolute temperature, R the universal gas constant, γ is a constant and $S(t_i)$ the cumulative stress drop. Since the seismic moment may be defined as $M_0 = \Delta\sigma \, V$, where V is the focal volume (Madariaga, 1979), the previous equation can be written as follow

$$M_{0m} \; + \; \sum_{j=1}^{k} M_{0j} \; = \; a \; + \; bLn(t_k) \tag{22}$$

where

$$a = V_k d\sigma \;\; ; \;\; b = \frac{V_k RT}{\gamma} \;\; ; \;\; V_k = V_0 + \sum_{j=1}^{k} V_j \tag{23}$$

V_k is the cumulative focal volume, V_0 is the focal volume of the main shock, V_j is the focal volume of the jth aftershock and M_{0m} and M_{0j} are the seismic moment of the main shock and the jth aftershock, respectively.

Following Marcellini (1997), we consider the aftershock zone as characterized by barriers that breaks after a given elapsed time proportional to the stress intensity factor K_i and therefore to the stress σ_i. Data fit of equation (22) characterizes the distribution, namely, the more our data explained by Eq. 22, the most likely the distribution of σ is close to the uniform distribution. This property is partially influenced by the spatial variation of the stress change produced by the mainshock, given that in the present static fatigue model a barrier breaks at σ_i, which is the superposition of the static stress before the mainshock and stress step induced by the mainshock In this study, as in Marcellini (1997), the seismic moment is evaluated using the Thatcher and Hanks (1973) relation, given by

$$Log M_0 = 9.0 + 1.5M \tag{24}$$

to test if the static model fatigue holds as represented by equation (22), two conditions must be checked:

(a) The validity of equation 22, which is adjusted to different aftershock data and plotted on Figure 8. The same figure shows the estimates a and b, their respective standard deviation σ_a and σ_b and the determination coefficient r^2 of the linear adjustment on the semi-logarithmic scale. Solid curves are plots of the function $y = a + bx$ where $x = \log(t)$. Dashed curves say (Δ_i^-), $y = a^- + b^- x$ and (Δ_i^+), $y = a^+ + b^+ x$, correspond to the 95% confidence limits relative to the confidence intervals $I(a) = [a^-,\ a^+]$ and $I(b) = [b^-,\ b^+]$ of a and b, respectively. Namely, $b^+ = \hat{b} + t_{n-2}^{(\alpha/2)} S_{\hat{b}}$, $b^- = \hat{b} - t_{n-2}^{(\alpha/2)} S_{\hat{b}}$; where $t_{n-2}^{(\alpha/2)}$ is the $\alpha/2$-quantile of the Student distribution with $n-2$ degrees of freedom ($\alpha = 0.05$ in our case) with

$$S_{\hat{b}} = \sqrt{\dfrac{\sum\limits_{i=1}^{n} \hat{\varepsilon}_i^2 \Big/ (n-2)}{\sum\limits_{i=1}^{n} \left(x_i - \bar{x}\right)^2}} \tag{25}$$

$\hat{\varepsilon}_i = y_i - \hat{y}_i$ are the regression errors and $\hat{y}_i = \hat{a} + \hat{b}x_i$ the predicted y_i values.

The confidence limit of $\hat{a}$ are calculated from those of $\hat{b}$ using the relation $\hat{a} = \bar{y} - \hat{b}\bar{x}$.

(b) The validity of the definition of the constants a and b as expressed previously by the relation (23). The conditions (a) and (b) are analyzed first by the quality of the fit, shown on figure 8.

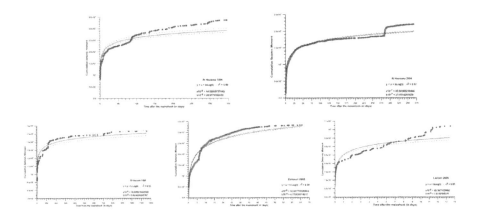

Figure 8. Fit of aftershock sequence, the plot displays the 95% confidence limit of the regression, the parameters of the adjustment and the coefficient of the correlation are shown on the plot. For Al Hoceima series of 1994 and 2004 the results are shown on graph (a) and (b) respectively. The graph (c) displays the results of the El Asnam serie of 10 october 1980. Graphs (d) and (e) display the results for the series of Zemouri 2003 and Laalam 2006.

Taking into account the obtained coefficient of correlation r^2 and given on each graph, the modele $y = a + b \, Log(t)$ introduced by Marcellini (1997) fit very well our data.

5. Energy partitioning

The other scaling law exalined in this study is the modified Bath law. In its original form, Bath law states that the difference Δm between a given mainshock with magnitude m_{ms} and its largest aftershock magnitude m_{as}^{max} is approximately constant, independently of the main-shock magnitude (Bath, 1965). That is,

$$\Delta m = m_{ms} - m_{as}^{max} \tag{26}$$

Δm is typically around 1.2.

Extensive studies about the statistical variability of Δm have been carried out by several authors e.g Vere-Jones (1969), Kisskinger and Jones (1991) and Console et al. (2003). However, the law remains an open problem nowadays (Vere-Jones, 1969; Vere-Jones et al. 2005). In this study, we use the modified Bath's law proposed by Shcherbakov and Turcote (2004a). It is based on an extrapolation of the Gutenberg-Richter relationship for aftershocks. Namely, the magnitude of the largest aftershock consistent with the Gutenberg-Richter relationship for aftershocks is obtained by formally putting $N(\geq m^*) = 1$ which yields to $a - bm^* = 0$. If Bath law is applicable to the inferres magnitude $m*$, the Gutenberg-Richter relationship takes the following form;

$$Log_{10}N(\geq m) = b\left(m_{ms} - \Delta m* - m\right) \tag{27}$$

where,

$$\Delta m* = m_{ms} - m_{as}^{*} \tag{28}$$

combining this last relation with the empirical relation of the energy dissipated

$$LogE = \alpha - \delta m \tag{29}$$

we derive the partionning of the energy, in the following way. It is wellestablished that the magnitude distribution of aftershocks clearly exhibits a near-universal scaling relative to the mainshock magnitude. To explore this relation for our aftershock sequences, we will determine the ratio of the total energy radiated by the afytershock sequence to the seismic energy radiated by the mainshock. The energy radiated during an earthquake is related empirically to its moment magnitude m by (Utsu, 2002)

$$\log_{10}E(m) = \frac{3}{2}m + \log_{10}E_0 \tag{30}$$

with,

$$E_0 = 6.3 \times 10^4 \, Joules \tag{31}$$

This relation is applied directly to describe the link between the energy radiated by the mainshock E_{ms} and the moment magnitude of the mainshock m_{ms},

$$E_{ms} = E_0 \cdot 10^{3/2 m_{ms}} \tag{32}$$

following Shcherbakov et al.(2004a), the total energy radiated during the aftershock sequence E_{as} is obtained by integrating over the distribution of aftershock till the inferred largest magnitude as upper bound in the integration. The upper bound of the integration is the largest aftershock magnitude unferred from the Gutenberg-Richter relationship and is denoted m^*. The total radiated energy in the aftershock sequence is obtained by integrating over the distribution of aftershock as,

$$E_{as} = \int_{-\infty}^{m^*} E(m)\left(-\frac{dN}{dm}\right) dm \tag{33}$$

m^* is the largest aftershock magnitude inferred from the Gutenberg-Richter law. Taking the derivative of the modified Bath law,

$$dN = -b(Ln10)10^{b(m_{ms}-\Delta m^*-m)}dm \tag{34}$$

by combining the former equations, Eq. 33 and 34, we obtain

$$E_{as} = b(Ln10)10^{b(m_{ms}-\Delta m^*)}\int_{-\infty}^{m^*} E(m)10^{-bm}\,dm \tag{35}$$

in addition, different version of the above equation, Eq 35 is obtained if we use the equation giving $E(m)$ as a function of E_0 and b value, we get

$$E_{as} = b(Ln10)10^{b(m_{ms}-\Delta m^*)}E_0\int_{-\infty}^{m^*} 10^{(\frac{3}{2}-b)m}\,dm \tag{36}$$

taking into account the modified Bath's law, we obtain

$$E_{as} = \frac{2b}{(3-2b)} E_0\, 10^{\frac{3}{2}(m_{ms}-\Delta m^*)} \tag{37}$$

the ratio of the total radiated energy by the aftershocks E_{as} to the total energy radiated by the mainshock E_{ms} is given by

$$\frac{E_{as}}{E_{ms}} = \frac{2b}{(3-2b)} 10^{-\frac{3}{2}\Delta m^*} \tag{38}$$

assuming that all earthquakes have the same seismic efficiency, which means that the ratio of the radiated energy to the total drop is stored as elastic energy is also the ratio of the drop in the stored elastic energy due to the aftershocks to the drop in the stored elastic energy due to the mainshock. Finally, the following relation is derived

$$\frac{E_{as}}{E_{ms} + E_{as}} = \left(1 + \frac{3 - 2b}{2b} 10^{3/2 \Delta m^*}\right)^{-1} \tag{39}$$

Using the Eq. 39 for the studied aftershock sequences, the results are shown on the following table.

	Ratio of the elastic energy released		
	$b \pm \sigma_b$	Δm^*	$E_{as}/{E_{as} + E_{as}}$
Al Hoceima 1994	1.07 ± 0.07	1.10	0.05
Al Hoceima 2004	1.13 ± 0.05	0.50	0.35
El Asnam 1980	0.82 ± 0.10	1.20	0.02
Zemouri 2003	1.10 ± 0.04	0.82	0.14
Laalam 2006	0.99 ± 0.10	1.70	0.01

Table 3. b-value, Bath law and energy partitionning derived for the different studied aftershock sequences.

From the obtained results shown on Table 3, we deduce the percentage of the total energy radiated during the mainshock. From this point of view, 95 % of the total energy has been radiated during the Al Hoceima 1994 mainshock and 65% during Al Hoceima 2004 mainshock. On the other side the El Asnam 1980 main shock radiated about 98% of the total energy, 2% has been radiated by the aftershock sequence, but we shouldpoint out that these results depend directly on the quality of data used and it is clear that whatever the sequence of aftershocks used, it is still incomplete, especially at the beginning just after the occurrence of the main shock. For the aftershock sequence triggered by the El Asnam earthquake of 1980, it seems that the sequence used is truncated due to the delay in the implementation of seismological network just after the main shock. Nevertheles, the results shown on Table 3, gives a large overview on the ratio of the total energy radiated by the main shock and by the aftershocks. Thus, 86 % of the total energy has been radiated during the main shock of Zemouri 2003 and 99% during Laalam 2006 mainshock.

6. Spatial aftershock distribution

It is well known that seismicity is a classical example of a complex phenomenon that can be quantified using fractal theory (Turcotte, 1997). In particular, fault networks and epicenter distributions have fractal properties (Goltz, 1998). Thus, a natural way to analyze the spatial distribution of seismicity is to determine the fractal dimension D_2. This D_2 - value is an extension of the Euclidian dimension and measures the degree of clustering of earthquakes. In the two dimensional space, D_2 can be a decimal number and ranges from 0 to 2.0. Therefore, the distributions characterized by different fractal dimensions defiine different clustering of events in space. In the two dimensional space, as D_2 approaches 1, the distribution of events approaches a line (Euclidean dimension equal to 1). The same occurs for the distribution of events along a fault. Alternatively, as D_2 approaches 2, the distribution of events tends to be uniform on the plane (Euclidean dimension equal to 2). When the D_2-values approaches 0, the distribution is concentrated in a single point (Beauval et al. 2006; Spada et al. 2011). In this study, the fractal dimension is estimated using the correlation dimension (Grassberger and Procaccia, 1983)

$$D_2 = \lim_{r \to 0} \frac{Log_{10} C(r)}{Log_{10}(r)} \tag{40}$$

where r is the radius of the sphere considered in the study, and $C(r)$ is the correlation integral given by

$$C(r) = \lim_{N \to +\infty} \frac{1}{N^2} \sum_{i=1}^{N} \sum_{j=1}^{N} H\left(r - |x_i - x_j|\right) \tag{41}$$

where N is the number of points in the analysis window, x_i; $i = 1,...., N$ are the coordinates of the epicenters, and $H(.)$ the Heaviside step function, $H(x) = 0$ for $x \leq 0$,

$H(x) = 1$ for $x \geq 0$. The quantity $C(r)$ is equivalent to the probability that two points will be separated by a distance less than r. The correlation integral is theoretically proportional to the power of D_2, i.e $C(r) \approx r^{D_2}$, where D_2 is the correlation dimension equals the second generalized Renyi dilmension d_2 (Molchan and Kronod, 2009). To estimate D_2 from the correlation integral (Spada et al. 2011), we plot $C(r)$ versus r on the log-log scale, Fig. 6., then we use the least square method to fit the data over the region where $C(r)$ is linear, which corresponds to the gradient of straight line of the resulting plot of $log_{10}C(r)$ against $log_{10}(r)$.

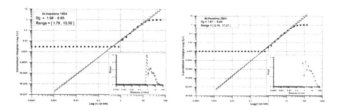

Figure 9. Graph displaying the plot of the correlation integrale in log-log scale, with the 95% of confidence limit (in dashed lines), the graphs show also, the plot the slope, related to the first derivative of the correlation integral for Al Hoceima 1994 and 2004 aftershock sequences.

In practice, however, for large values of r the gradient is artificially low, whereas for small values of r the gradient is artificially high. These two cases have been called "saturation" and "depopulation" (Nerenberg and Essex, 1990). Whereas, it is common in the estimation of the fractal dimension to use fiting procedure to straight line than to a subjectivelly chosen straight part of the curve. Nerenberg and Essex (1990), give formulas for determining the distances of depopulation and saturation, r_d and r_s given by

$$r_d = 2R\left(\frac{1}{N}\right)^{1/d} \quad and \quad r_s = \frac{R}{d+1} \tag{42}$$

where d is the dimensionality of the data cluster, and $2R$ is the approximate lenght of the side of the area containing data. As pointed by Eneva (1996), it is safe to start the scaling range at values of r as low as $r_d / 3$, but in this study, we have use the slope method, which as explained previously consist in estimating the slope of the double logarithmic plot of the correlation integral versus distance. The stability of the scaling range is verified with the slope. It consists in calculating the first derivative between each two pints of the correlation integrale curve and plotting versus logarithmic distance. To eliminate the depopulation and saturation effects, the scaling range is defined within the region where the slope is most constant. Using this procedure and as for Al Hoceima 1994 and 2004 series, the results obtained are displayed in the following figure

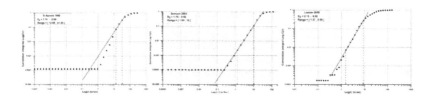

Figure 10. Graph showing the correlation integral $Log_{10}C(r)$ vs $Log_{10}(r)$.

The results obtained are close to 2.0, which allow us to deduce that the spatial distribution of the epicenters tends to be uniform on the plane.

The ratio of the slip on the primary fault to the total slip over the fault system is given by (Khattri, 1995)

$$\frac{S_p}{S} = 1 - 2^{-(3-D_2)} \tag{43}$$

where S_p is the slip over primary fault and S represents the total slip over the fault-system.

The obtained results are shown on Table 3, we deduce that during Al Hoceima earthquake of 1994, 62% of the total slip accomodates the primary rupture, 60 % during Al Hoceima 2004 earthquake. During the El Asnam earthquake of 10 October 1980, 59% of the total slip accomodates the primary fault segment, during Zemouri earthquake of 2003 this ratio has been estimate to 57% and 45% during Laalam earthqiuake of 2006. It is important to point out that in each case the remainder of the slip is distributed over the secondary rupture.

	Fractale dimension and ratio of the slip		
	$D_2 \pm \sigma_{D_2}$	Range	S_p/S
Al Hoceima 1994	1.60 ± 0.05	1.79 - 12.56	0.62
Al Hoceima 2004	1.67 ± 0.04	2.16 - 17.27	0.60
El Asnam 1980	1.70 ± 0.09	12.90 - 31.35	0.59
Zemouri 2003	1.79 ± 0.02	1.60 - 10.00	0.57
Laalam 2006	2.13 ± 0.02	1.57 - 8.90	0.45

Table 4. Fractale dimension D_2 and ratio of the slip on the primary fault over the fault system.

In this section we attempt to analysis the inter-event distance distribution of probability. We use a non-parametric approach to analysis the density of probability of the inter-event distances, especially the kernel density estimation, this methodlogy is clearly presented in Silverman (1986). We used it in the following way. Given a sample of n observations $x_1, x_2, \ldots, x_n$ with unknown probability distribution function f, the kernel density estimate of f is given by,

$$\hat{f}(x,h) = \frac{1}{nh}\sum_{i=1}^{i=n} K\left(\frac{x-x_i}{h}\right) \tag{44}$$

where $K(.)$ is a positive function called kernel, a typical example is the Gaussian kernel defined by;

$$K(x) \quad = \quad \frac{1}{\sqrt{2\pi}} e^{-\frac{1}{2}x^2} \tag{45}$$

in Eq. 44, h is a parameter and its determination is crucial. There are differents method to estimate the parameter h, in the following lines we summarise the least-square cross-validation introduced by Silverman (1986, pp 48). The kernel density estimate of f is given as shown previously by

$$\hat{f}(x,h) \quad = \quad \frac{1}{nh}\sum_{i=1}^{i=n}K\left(\frac{x-x_i}{h}\right) \tag{46}$$

considere a sample of n observations $x_1, x_2, \ldots\ldots, x_n$; the quadratic uncertainty is then given by

$$\int_{-\infty}^{+\infty}\left[\hat{f}(x)-f(x)\right]^2 dx \quad = \quad \int_{-\infty}^{+\infty}\left[\hat{f}(x)\right]^2 dx - 2\int_{-\infty}^{+\infty}\left[f(x)\hat{f}(x)\right]dx + \int_{-\infty}^{+\infty}\left[f(x)\right]^2 dx \tag{47}$$

The last term of the right member of the last equality is independnat of the parameter h, thus the optimal value of h is obtained by minimizing the two others terms. The problem then consists to find h minimizind the score function defined as follow

$$M_0(h) = \int_{-\infty}^{+\infty}\left[\hat{f}(x)\right]^2 dx \;-\; \frac{1}{2n}\sum_{i=1}^{n}\hat{f}_{-i}(x_i) \tag{48}$$

where,

$$\hat{f}_{-i}(x_i) \quad = \quad \frac{1}{(n-1)h}\sum_{i\neq j}K\left(\frac{x-x_i}{h}\right) \tag{49}$$

futhermore, the score function could be written in the following form

$$M_0(h) \quad = \quad \frac{1}{n^2h}\sum_{i=1}^{n}\sum_{j=1}^{n}K^{(2)}\left(\frac{x_i-x_j}{h}\right) \;-\; \frac{2}{n(n-1)h}\sum_{i=1}^{n}\sum_{j=1}^{n}K\left(\frac{x_i-x_j}{h}\right) \tag{50}$$

with,

$$K^{(2)}(x) = \int_{-\infty}^{+\infty} K(\xi).K(x-\xi)d\xi \tag{51}$$

assuming that the minimum of $M_0(h)$ is in the vicinity of the minimum of $E[M_0(h)]$ and then

it is in the vicinity of $E\left[\int_{-\infty}^{+\infty}[\hat{f}(x) - f(x)]^2 dx\right]$. Also, for n large, $(n-1) \cong n$; finaly the problem

consites to minimize the simplified score function defined as follow, we replace $M_0(h)$ by $M_1(h)$ given by

$$M_1(h) = \frac{1}{n^2h^2}\sum_{i=1}^{i=n}\sum_{j=1}^{j=n}\left[K^{(2)}\left(\frac{x_i - y_j}{h}\right) - 2K\left(\frac{x - x_i}{h}\right)\right] + \frac{2}{nh}K(0) \tag{52}$$

we observe that in the case of the Gaussian kernel, the kernel $K^{(2)}(.)$ is a centered Gaussian distribution with variance 2. In this case the simplified score function is written as follow

$$M_1(h) = \frac{1}{n^2h^2}\frac{1}{\sqrt{\pi}}\sum_{i,j}\left[\frac{1}{2}\exp\left\{\frac{(x_i-y_j)^2}{4h^2}\right\} - \sqrt{2}\exp\left\{\frac{(x_i-y_j)^2}{2h^2}\right\}\right] + \frac{\sqrt{2}}{nh\sqrt{\pi}} \tag{53}$$

the parameter h minimizing the simplified score function is then obtained as solution of the following equation

$$\sum_{i,j}\left[\frac{1}{\sqrt{2}}\left\{\frac{(x_i - x_j)^2}{2h^2} - 1\right\}\exp\left\{-\frac{(x_i - x_j)^2}{4h^2}\right\} - 2\left\{\frac{(x_i - x_j)^2}{h^2} - 1\right\}\right] - 2n = 0 \tag{54}$$

the parameter h obtained using teh cross-validation method is then given by

$$h_{cv} = Arg\max\left(\prod_{i=1}^{n}\hat{f}_{-i}(x_i,h)\right) \tag{55}$$

where $\hat{f}_{-i}(., h)$ is the kernel density calculated from the sample $x_1, x_2, \ldots, x_{i-1}, x_{i+1}, \ldots, x_n$. We use the Silverman multi-modal tests to estimate the number of true bumps. This test is as follow $T^{(k)}$ test the null hypothesis $H_0^k:$ "f has k bumps" against the alternative H_1^k: " f has more than k bumps". Under the hypothesis H_0^k, the smoothing parameter of the Gaussian kernel density estimate $\hat{f}(., h)$ is given by

$$h_{crit}(k) = Inf\left\{h \in \mathbb{R} \ / \ \hat{f}(.,h) \text{ has } k \text{ bumps or less}\right\} \tag{56}$$

The test $T^{(k)}$ is inspired from the fact that big values of parameter $h_{crit}(k)$ reject the null hypothesis H_0^k. The following therorem by Silverman (1981) gives a characterisation

Theorem (Silverman, 1981)

The kernel density estimate $\hat{f}(., h)$ with Gaussian kernel, has more than k bumps if and only if $h < h_{crit}(k)$.

The test $T^{(k)}$ is constructed by simulating N statistics $h_{crit}^{(1)}(k) ; h_{crit}^{(2)}(k) ; h_{crit}^{(3)}(k); ....; h_{crit}^{(N)}(k)$ from N smoothed bootstrap samples of size n obtained from $x_1, x_2, ..., x_n$.

Proof. The proof of this theorem is given in details in Silverman (1981).

Under the null hypothesis, samples can be simulated from the kernel density estimate by using Efron formula,

$$y_i = \bar{x} + \left(1 + \frac{h^2}{\sigma^2}\right)^{-\frac{1}{2}}\left(x_{I(i)} - \bar{x} + h_{crit}(k)\varepsilon_i\right) \tag{57}$$

where $(x_{I(i)}; i=1,n)$ is a bootstrap sample of size n simulated from the sample $x = (x_1, x_2, ..., x_n)$; $\bar{x}$ and σ^2 are the mean value and variance of the sample x, and $(\varepsilon_i; i=1,n)$ is a randomly simulated sample from the standard normal distribution.

if $h_{crit}^*(k)$ is the parameter obtained from x then the p value of this test is given by

$$P_{value} = \frac{\#\left\{h_{crit}^{(i)}(k) > h_{crit}^*(k)\right\}}{N} \tag{58}$$

is the sign of the number of element in the set. In practice, we apply the series of tests $T^{(1)}, T^{(2)}, T^{(3)}, ...,$ in this order until the null hypothesis is not rejected. We used the previous methods, to derive the distribution of the inter-event distances. The results obtained for the aftershock sequence triggered by the Al Hoceima 1994 earthquake are shown on Fig. 11.

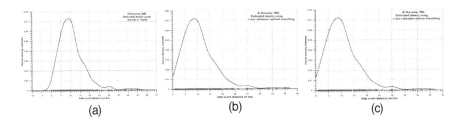

Figure 11. Kernel density estimated for Al Hoceima 1994 aftershock sequence (a) using the rule of thumb (b)using cross validation smoothing and (c) Estimated density obtained with one mode critical bandwidth h_{crit} = 6.4

The Fig. 11(a) gives the estimated density obtained using the rule of thumb. The cross validation optimal smoothing method gives a bandwidth parameter h = 1.2, the estimated density
of probability obtained is shown on Fig. 11(b). The estimated density obtained with one mode
critical bandwidth equal to h_{crit} = 6.4 is shown on Fig. 11(c). On the three Figures a concentration of the inter-event distances around the value 8.9 km, which correspond to the mode of the
estimated distribution of probability. The results obtained for the Al Hoceima 2004 aftershock
sequence are shown on the Fig. 12. The estimated density using the rule of thumb shown on
Fig 12(a) displays a concentration of the inter-event distances around the value 9.45 km, the
estimated density using the cross validation optimal smoothing has been obtained using a
bandwidth parameter h = 1.0. The density obtained displays a concentration of the inetr-event
distances around the value 7.5 km. Although, the estimated density with one mode critical
bandwidth obtained equal to h_{crit} = 11, shows a concentration of the inter-event distances
around the value 14.14 km.

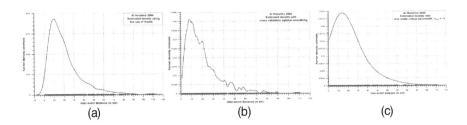

Figure 12. Kernel density estimated for Al Hoceima 2004 aftershock sequence (a) using the rule of thumb (b)using cross validation smoothing and (c) Estimated density obtained with one mode critical bandwidth h_{crit} = 11

For the aftershock sequence triggered by the 21 May 2003 Zemouri earthquake (Mw 6.9), the
obtained results are shown on Fig. 13.

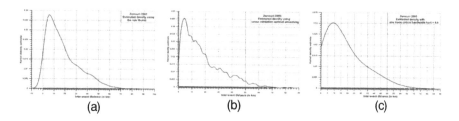

Figure 13. Kernel density estimated for Zemouri 2003 aftershock sequence (a) using the rule of thumb. (b) using cross validation smoothing and (c) Estimated density obtained with one mode critical bandwidth $h_{crit} = 6.6$

The estimated density using the cross validation optimal smoothing has been obtained with a bandwidth parameter $h = 1.0$. The density displays an inter-event distances concentration around the value 4.29 km as shown on Fig. 13(b). The estimated density with one mode critical bandwidth $h_{crit} = 6.6$; displays a concentration of the inter-event distances around 9.0 km.

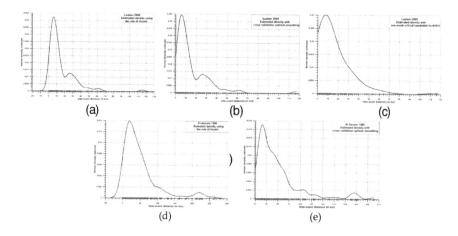

Figure 14. Kernel density estimate for Laalam 2006 aftershock sequence ((a), (b) and (c)). The Graphs on (d) and (e) display the kernel density estimate for El Asnam 1980 aftershock sequence.

7. Conclusion

Aftershock sequences in Algeria-Morocco region have been analyzed in order to estimate and derive with accuracy the parameters of the most important scaling laws in statistical seismology. For each aftershock sequence the threshold complteness magnitude m_c has been derived using different approach, we have choose the one giving the "most" stability of the fit of the cumulative number by a straight line. For the magnitude above the threshold magnitude above the threshold completeness m_c, we used the maximum likelihood to estimate above the threshold completeness m_c. The results obtained are close to 1.0, the typical universal value for aftershock sequence. The Omori-Utsu law for aftershock decay and Bath's law for the difference between magnitude and the largest aftershock and mainshock as modified by Shcherbakov et al., (2004a, b)., while using *AIC* as a measure to select the most appropriate model between the first stage and two stage Omori-Utsu model, the rate of decay of aftershocks was found to follow in all case, except the May 21, 2003 Zemouri aftershock sequence, the first stage Omori-Utsu law, which mean the Omori-Utsu model without secondary aftershock. The decay of aftershock trigged by the May 21, 2003 Zemouri earthquake exhibit a better fit with a two stage Omori-Utsu model, denoted model 4 in the text. The later allow us to include in the modelisation the secondary aftershock of magnitude 5.8 m_{bLg}. The study of the Omori-Utsu law has been complemented of aftershock sequences. This procedure is an alternative to the Omotri-Utsu law based on the physics of the static fatigue mechanism. The study of Gutenberg-Richter relationship and the modified Bath's law, allowed us to perform the partitioning of the energy released during each aftershock sequence by the main shock and aftershocks. The spatial analysis has been performed using the fractal dimension D_2 derived using the Correlation integrale. Using non parametric estimation approach, especially the kernel density estimation, we derive the estimation of the density of the probability of the inter-event distances. The mode of this density highlight how the spatial clusters of the events.

This study is a first attemp to perform analysis of aftershock sequneces triggered by main events in the Algeria-Morocco region.

Author details

M. Hamdache[1], J.A. Peláez[2] and A. Talbi[1]

1 Seismological Survey Department. C.R.A.A.G. Algiers, Algeria

2 Department of Physics, University of Jaén, Jaén, Spain

References

[1] Akaike, H. (1974). A new look at the statistical model identification, IEEE Trans. Autom. Control , 19, 716-723.

[2] Aki, K. (1965). Maximum likelihood estimate of b in teh formula Log N = a- bM and its confidence limits,. Bull. Earthq. Res. Inst. Tokyo Univ. , 43, 237-239.

[3] Armorese, D. (2007). Applying a change of point detection method on frequency-magnitude distributions.Bull. Seismology Society of America, doi.

[4] Bath, M. (1965). Lateral inhomogeneties in the upper mantle. Tectonophysics, , 2, 483-514.

[5] Beauval, C, Hainzl, S, & Scherbaum, F. (2006). The Impact of the spatial uniform distribution of seismicity on probabilistic seismic-hazard estimation. Bulletin Seismology Society of America. n6 , 96, 2465-2471.

[6] Beldjoudi, H, Guemache, A, Kherroubi, A, Semmane, F, Yelles-chaouche, K. A, Djellit, H, Amrani, A, & Haned, A. (2009). The Laalam (Bejaia, Nort-East Algeria) Moderate earthquake (Mw=5.2) on March 20, 2006. Pure and Applied Geoph. DOIs00024-009-0462-9.

[7] Bender, B. (1983). Maximum-Likelihood estimation of b-values for magnitude grouped data, Bull. Seismol. Soc. Am. 73, n°, 3, 831-851.

[8] Calvert, A, Gomez, F, Seber, D, Baranzagi, M, Jabour, N, Ibenbrahim, A, & Demnati, A. (1997). An integrated geophysical investigation of recent seismicity in the Al Hoceima region of North Morocco. Bulletin Seism. Soc. of Am. 87. , 637-651.

[9] Console, R, Lombardi, A. M, Murru, M, & Rhodes, R. (2003). Bath's law and the self-similarity. J. Geophys. Res. 108, 2128.

[10] Daley and Vere-Jones(2003). An introduction to the theory of point process, 2nd Ed., Springer-verlag. New-York. Berlin Heidelberg., 1

[11] El Alami, S. O, Tadili, B. A, Cherkaoui, T. E, Medina, F, Ramdani, M, Ait-brahim, L, & Hanafi, M. (1998). The Al Hoceima earthquake of May 26, 1994 and its aftershocks : a seismotectonic study. Annali di Geofisica, , 41, 519-537.

[12] Eneva, M. (1996). Effect of limited data sets in evaluating the scaling properties of spatially distributed data : an example from minning-induced seismic activity, Geophys., J. Int., , 124, 773-786.

[13] Enescu, B, & Ito, K. (2002). Spatial analysis of the frequency-magnitude distribution and decay rate of aftershock activity of the 2000 Western Tottori earthquake. Earth Planets Space , 54, 847-859.

[14] Enescu, B, Mori, J, Masatoshi, M, & Kano, Y. (2009). Omori-Utsu law c-values associated with recent moderate earthquakes in Japan, Bull. Seismol. Soc. of Am. 99, 2A, , 884-891.

[15] Frohlich, C, & Davis, S. D. (1990). Single-link cluster analysis as a method to evaluate spatial and temporalproperties of earthquake catalogues. Geophy. J. Inter. , 100, 19-32.

[16] Galindo-zaldivar, J, Chalouan, A, Azzouz, O, San, C, De Galdeano, F, Anahnah, L, Ameza, P, Ruano, A, Pedrera, A, Ruiz-constan, C, Marin-lechado, M, Benmakhlouf, A. C, Lopez-garrido, M, Ahmamou, R, Saji, F. J, & Roldan-garcia, M. Akli., and A.

Chabli. (2009). Are seismological and geological observations of the Al Hoceima (Morocco Rif) 2004 earthquake (M=6.3) contradictory. Tectonophysics,

[17] Gardner, J. K, & Knopoff, L. (1974). Is the sequence of earthquake in southern California, with aftershock removed, Poissonian ? Bull. Seism. Soc. Am., 64 (5), 1363-1367.

[18] Goltz, C. (1998). Fractal and chaotic properties of earthquake, in Lecture Notes in Earth Sciences, Springer, New York, 175 pp.

[19] Grassberger, P, & Procaccia, I. (1983). Measuring the strangeness of strange attractors. Physics D, , 9, 189-208.

[20] Gutenberg, R, & Richter, C. F. (1944). Frequency of Earthquake in California. Bull. Seismol. Soc. Am. 34. , 158-188.

[21] Guo, Z, & Ogata, Y. (1997). Statistical relation between the parameters of aftershocks in time, space and magnitude. Journal of Geophys. Res. 102(B2)., 2857-2873.

[22] Guttorp, P, & Hopkins, D. (1986). On estimating varying b-values, Bull. Seismol. Soc. Am. 76 n°, 3, 889-895.

[23] Hamdache, M, & Pelaéz, J. A. and K. Yelles Chaouche. (2004). The Algiers, Algeria earthquake (Mw 6.8) of the 21 May 2003: Preliminary report. Seism. Res. Letters, n3. May/June 2004., 75

[24] Hamdache, M, Pelaéz, J. A, & Talbi, A. and López Casado, C. (2010). A unified catalog of main earthquakes for Northern Algeria from A.D. 856 to 2008. Seism. Res. Lett. , 81, 732-739.

[25] Kagan, Y. Y. (2004). Short-term proprieties of earthquake catalogs and models of earthquake source. Bull. Seismol. Soc. Am. 94 (4), 1207-1228.

[26] Khattri, K. N. (1995). Fractal description of seismicity of India and inferences regarding earthquake hazard. Curr. Sci., 69. , 361-366.

[27] Kisslinger, C, & Jones, L. M. (1991). Proprieties of Aftershocks in Southern California. J. Geophy. Res. , 103(24), 453-24.

[28] Maouche, S, Meghraoui, M, Morhange, C, Belabbes, S, Bouhadad, Y, & Haddoum, H. (2011). Active coastal thrusting and folding, and uplift rate of the Sahel Anticline and Zemouri earthquake area (Tell Atlas, Algeria). Tectonophysics, 509, , 69-80.

[29] Marcellini, A. (1995). Arrhenius behavior of aftershock sequences. J. Geophys. Res. , 100, 6463-6468.

[30] Marcellini, A. (1997). Physical model of aftershock temporal behavior. Tectonophysics , 277, 137-146.

[31] Marzocchi, W, & Sandri, L. and new insights on the estimation of the b-value and its uncertainty. Annals of Geophysics. n6., 46

[32] Mikumo, T, & Miyatake, T. (1979). Earthquake sequences on a frictional fault model with non-uniform strenghs and relaxation times, Geophys. Journal R. Astron. Soc., , 59, 497-522.

[33] Mogi, K. (1962). Study of elastic shocks caused by the fracture of heterogeneous materials and its relation to the earthquake phenomena, Bull., Earthquake Res., Inst., Univ., Tokyo, , 40, 125-173.

[34] Molchan, G, & Kronod, T. (2009). The fractal description of seismicity, Geophs. J. Int. 179. n doij.1365., 3, 1787-1799.

[35] Nerenberg, M. A. H, & Essex, C. (1990). Correlation dimension and systematic geometric effects, Phys. Rev. A. 42, , 7065-7074.

[36] Nocquet, J. M, & Calais, E. (2003). Crustal velocity field of Western Europe from permanent GPS array solutions, 1996-2001. Geoph. J. International, , 154, 72-88.

[37] Nyffengger, P, & Frolich, C. (1998). Recommandations for determining p values for aftershock sequence and catalogs. Bull. Seismol, Soc. Am. n0 5 , 88, 1144-1154.

[38] Nyffengger, P, & Frolich, C. (2000). Aftershock occurrence rate decay properties for intermediate and deep earthquake sequences. Geoph. Res. Lett. , 27, 1215-1218.

[39] Ogata, Y. (1983). Estimation of the parameters in the modified Omori formula for aftershock frequencies by the maximum likelihood procedure. J. Phys. Earth. , 31, 115-124.

[40] Ogata, Y. (1992). Detection of precursory relative quiescence before great earthquake through a statistical model. Geophys. Res. , 97(19), 845-19.

[41] Ogata, Y, & Katsura, K. (1993). Analysis of temporal and spatial heterogeinity of magnitude frequencey distribution inferred from earthquake catalogue. Geophys. J. Int. 113, , 727-738.

[42] Ogata, Y. (1999). Seismicity Analysis through Point-Process Modelling : A Review. Pure and Applied Geophysics, 155. , 471-507.

[43] Ogata, Y, Jones, L. M, & Toda, S. (2003). When and where the aftershock activity was depressed : Contrasting decay patterns of the `proximate large earthquake in southern California. J. of Geophysics Research, n0 B6 2318, doiJB002009., 108

[44] Olssen, R. (1999). An estimation of the maximum b values in the Gutenberg-Richter relation. Geodynamics, , 27, 547-552.

[45] Omori, F. (1894). On the aftershocks of earthquake. J. Coll. Sci. Imp. Univ. Tokyo, 7. , 111-120.

[46] Ouyed, M, Meghraoui, M, Cisternas, A, Deschamp, A, Dorel, A, Frechet, F, Gaulon, R, Hatzfeld, D, & Philip, H. (1981). Nature, , 292(5818), 26-31.

[47] Peláez, J. A, Chourak, M, Tadili, B. A, Brahim, L. A, & Hamdache, M. López Casado, C., and Martínez Solares, J.M. (2007). A catalog of main Moroccan earthquakes from 1045 to 2005. Seismological Research Letters , 78, 614-621.

[48] Peláez, J.A., M. Hamdache and Sanz De Galdeano. 2012. A spatially smoothed seismicity forecasting model for Mw≥5.0earthquakes in northern Algeria and Morocco. 15 World Conf. Earth. Eng. 24 - 28 September 2012, Lisboa, Portugal

[49] Reasenberg, P. (1985). Second-order moment of central California seismicity, 1969-82, J. Geophys. Res., , 90, 5479-5495.

[50] Rydelek, P. A, & Sacks, I. S. (1989). Testing the completeness of earthquake catalogues and the hypothesis of self-similarity, Nature, , 337, 251-253.

[51] Sandri, L, & Marzocchi, W. (2005). A technical note on the bias in the estimation of the b-value and its uncertainty through the least squares technique. Annals of Geophysics.

[52] Sanz de GaldeanoC., (1990). Geologic evolution of the Betic cordilleras in the Western Mediterranean, Miocene to Present. Tectonophysics , 172, 107-119.

[53] Shcherbakov, R, & Turcotte, D. L. (2004a). A modified form of Bath's law. Bull. Seismol. Soc. Am. , 94, 1968-1975.

[54] Shcherbakov, R, Turcotte, D. L, & Rundle, J. E. (2004b). A generalized Omori´s law for earthquake aftershock decay. Geophy. Res. Lett. 31. L11613, doiGL019808.

[55] Shcherbakov, R, Turcotte, D. L, & Rundle, J. E. (2005). Aftershocks Statistics. Pure and Applied Geophysics , 162, 1051-1076.

[56] Shcherbakov, R, & Turcotte, D. L. (2006). Scaling properties of the Park-field aftershock sequence. Bull. Seismol, Soc. Am. 94. SS384., 376.

[57] Shi, Y, & Bolt, B. (1982). The standard error of the magnitude-frequency b value. Bull. Seism. Soc. Am. n°5 , 72, 1677-1687.

[58] Silverman, B. W. (1986). Density estimation for statistics and data analysis. Chapman and Hall, London.

[59] Spada, M, Weimer, S, & Kissling, E. (2010). Quantifying a potential bias in probabilistic seismic hazard assessment: Seismotectonic zonation with fractal properties. Bull. Seism. Soc. Am. n 6 , 101, 2694-2711.

[60] Stich, D, Mancilla, F, Baumont, D, & Morales, J. (2005). Source analysis of the Mw 6.3 2004 Al Hoceima earthquake (Morocco) using regional apparent source time functions. J. Geoph. Res. 110 (B06306) doiJB003366.

[61] Tahir, M, Grasso, J. R, & Amorèse, D. (2012). The largest aftershock: How strong, how far away, how delayed?. Geophysical Research letters, L04301, doi:GL050604, 2012., 3

[62] Thatcher, W, & Hanks, T. C. (1973). Source parameters of southern California earthquake. J. Geophys. Res. , 78, 8547-8576.

[63] Tinti, S, & Mulargia, F. (1987). Confidence intervals of b-values for gropuped magnitudes. Bull. Seismol. Soc. Am. 77, n°, 6, 2125-2134.

[64] Turcotte, D. (1997). Fractals and chaos in Geology and Geophysics, Cambridge University Press, New York, 416 pp

[65] Utsu, T. (1961). A statistical study on the occurrence of aftershocks. Geophysics, 30. , 521-605.

[66] Utsu, T. (1965). A method for determining the value of b in a formula log n = a- bM showing the magnitude-frequency relation for earthquakles, Geophys, Bull. Hokkaido Univ. , 13, 99-103.

[67] Utsu, T. (1969). Aftershocks and earhquake statistics (I)- Some Parameters which characterize an Aftershock Sequence and their Interaction. Journal Fac., Sci., Hokaido, Univ., Ser., VII (Geophys.), , 3, 129-195.

[68] Utsu, T. (1971). Aftershocks and Earthquake Statistics (III). Analyses of teh distribution of earthquakes in magnitude, Time and Space with special consideration to clustering characteristics of earthquake occurrence. Jpurnal of the Faculty of Science, Hokkaido Univ. Ser. VII, Geophysics, Vol. III, n°5.

[69] Utsu, T, Ogata, Y, & Matsu, R. S. ra. (1995). The centenary of the Omori formula for a decay law of aftershock activity. J. Phys. Earth 43. , 1-33.

[70] Vere-jones, D. (1969). A note on the statistical interpretation of Bath`s law. Bulletin Seismol. Soc. of Am. n4, , 69, 1535-1541.

[71] Vere-jones, D, Murakami, J, & Christophersen, A. (2005). A further note on Bath's law, The 4th International Workshop on Statistical Seismology.Tokyo. Japan.

[72] Vidal, F. (1986). Sismotectónica de la región Béticas-Mar de Alborán. PhD Thesis. Universidad de Granada. 457 p.

[73] Wiemer, S, & Zuniga, R. F. (1994). Zmap. A Software package to analyse seismicity (abstract), EOS Trans. AGU 75 (43), Fall Meet. Suppl., 456.

[74] Wiemer, S, & Katsumata, K. (1999). Spatial variability of seismicity parameters in aftershock zones. J. Geophys. Res. 104 (B6), 13, 135-151.

[75] Wiemer, S, & Wyss, M. (2000). Minimum magnitude of completness in earthquake catalogs : Examples from Alaska, the Western United States, and Japan. Bull. Seismol. Soc. Am. , 90, 859-869.

[76] Wiemer, S. (2001). A software package to analyze seismicity: ZMAP, Seismol. Res. Lett., , 72, 374-383.

[77] Wiemer, S, & Baer, M. (2000). Mapping and removing quarry blast events from seismicity catalogs, Bull. Seism. Soc. Am., , 90, 525-530.

[78] Wiemer, S, & Wyss, M. (2000). Minimum magnitude of complete reporting in earthquake catalogs: examples from Alaska, the Western United States, and Japan, Bull. Seism. Soc. Am. , 90, 859-869.

[79] Woessner, J, & Wiemer, S. (2005). Assessing the Quality of Earthquake Catalogues: Estimating the Magnitude of Completeness and Its Uncertainty, Bull. Seismol. Soc. Am., doi:10.1785/0120040007,, 95, 684-698.

Chapter 3

Global Climatic Changes, a Possible Cause of the Recent Increasing Trend of Earthquakes Since the 90's and Subsequent Lessons Learnt

Septimius Mara and Serban-Nicolae Vlad

Additional information is available at the end of the chapter

1. Introduction

Over 1 million earthquakes a year can be felt by people on Earth. Large earthquakes and related effects rank among most catastrophic environmental events. Both tectonically active areas of lithospheric plates interactions along their boundaries and intra-plate fault displacements are responsible for rupture yielding seismic waves that shake the ground. Devastating effects of the earthquakes that occurred during the last decades underlines the necessity of a multi-hazard approach regarding the subsequent effect of the tremor waves, such as tsunami waves (Sumatra – Andaman Islands, 2004, NE Japan 2011), soil subsidence and major accidents at nearby chemical facilities (Turkye, Kocaeli, 1999), explosions at petrochemical and nuclear plant, after failure of the cooling system due to power failure, following the 10 meters tsunami wave (NE Japan 2011), submarine landslides (northern coast of Papua New Guinea, 1998), or landslides and soil liquefaction ("earthquake lake" at Sichuan, China, 2008, Christchurch, New Zeeland, 2011). The multi-hazard concept represents a new direction of research in an integrated manner, with applied global implications. The frequency of the disasters appears to increase in the last decades (Fig. 1,2), and the communities became more vulnerable to the natural hazards, generally due to the complex aspects generated by increased urbanization, land planning and environmental changes. The uncertainties involving the relations between different components of the surrounding environment made more difficult the investigation of each category of natural hazards [1]. Consequently it is necessary to study groups of hazards, not just a single case, and the interaction among them in order to have a clear view of the internal processes and causative factors of the disasters. From this point of view, the disaster seems to be more internationalized, due to global factors which interact and affect the population and the environmental factors.

2. Problem statement

Recently it became relevant that, despite frequent large earthquakes, several countries located in prone areas didn't have strong building codes and many houses are built out of mud bricks and un-reinforced masonry, which do not stand up well to earthquakes. Mud brick didn't resist to the earthquake stress and too heavy tile and cement roofs generally collapsed into many houses. Other factors contribute to the severity of a quake, but earthquake resistant buildings can make a huge difference in the number of damages [2]. As a result, casualties and damage are much higher than similar earthquakes elsewhere in the world. Therefore recent major earthquakes such as Guarajat, India (2001), Bam-Iran (2003), Sumatra – Andaman Islands (2004), Kashmir-Pakistan (2005), South of Java – Indonesia (2006) or Sichuan, China (2008) led to heavy human casualties, compared with other similar earthquakes all over the world. The same magnitude earthquakes, for example the Northridge quake in Los Angeles in 1994 killed only 57 people and in Kobe Japan in 1995 a similar quake killed about 5,000. Another example could be the earthquake –magnitude 7 - from Haiti, at Port-au-Prince in January 2010, with almost 220,000 casualties compared with a similar earthquake in the next month, in Chile, magnitude 8.8, 500 times higher than the previous one in Haiti, resulted in less than 600 casualties. In case of major tsunamis, which cross an entire Ocean, or so called "tele-tsunamis", i.e the greater earthquake ever recorded by instruments, with a 9.5 magnitude, in Valvidia, Chile (1960), which produced damage in Hawaii and alarm in Japan, it became obviously the "globalisation" of the subsequent effects of the tremors. They can reach any coastal areas all over the world, not necessarily earthquake prone areas, and request dedicated building codes. A similar effect took place following the recent great earthquakes at Sumatra – Andaman Islands (2004), 9.1 magnitude, with damages 1 mile inside the affected coastal areas, with a maximum height of the tsunami wave up to 30m, or the recent NE of Japan (2011), magnitude 9, where tsunami waves inflicted severe damages 9 miles inside the coast areas. The recent catastrophe in Japan exceeds the worst case scenarios previously estimated in prevention measures, especially at the nuclear plants. The maximum possible height of a tsunami wave was estimated at 6 meters high, whereas the height of the wave reached 10 m (the maximum recorded height was 23m for the NE of Japan).

3. Application area

The present analysis is based on data regarding the earthquake frequency and magnitude the world over, (Fig. 2), i.e. USGS (United States Geological Service) data base during the last 30 years [4]. It has to be specified that the earthquake monitoring activity network was used during the cold war [3], since 50's, to identify and localise nuclear tests all over the world, taking into account that a nuclear detonation is detected generally less than a 6 magnitude earthquake on Richter scale, where is produced, depending of the distance from the source. Therefore the data taken into consideration in the present earthquake evaluation includes just important earthquakes, which can produce significant damage (above level of stronger earthquakes, with the magnitude over 6 on Richter scale). As a conclusion, the thesis stipulating that just in the recent year the global network of seismographs was completed and that's why we have the "felling" of an increased trend of the earthquakes, and therefore the study of the past earthquake data didn't reflect complete the reality, because of "missing" earthquake, is falls. According with this theory, the same increasing pattern of the earthquakes should be observed since 80's, but as observed in the evolution trend of the similar earthquake magnitudes over each separate decade (in 80's and 90's), are significantly different, in both of magnitudes range and increasing trend from one year to another (please see below Fig.1.a-d, of evolution including subsequent linear trends). Should be noticed that only for the decade 1980-1990, the trend line is decreasing, compared with the period intervals of 1990-2000 and 2000-2010, when the evolution trend of earthquake frequency and magnitude, in visible increasing.

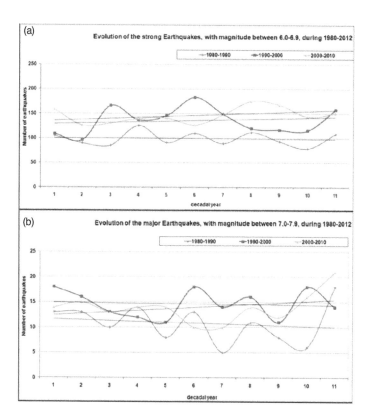

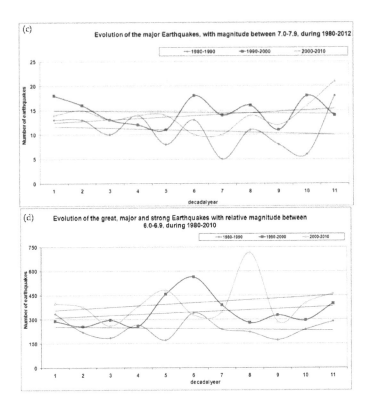

Figure 1. a. The strong earthquake type of 6 - 6.9 magnitude on Richter scale:b. The major earthquake type with the magnitude of 7 - 7.9 magnitude on Richter scale:c. The great earthquake type, with the magnitude over 8 magnitude on Richter scale:d. The great, major and strong earthquakes types, with the relative magnitude of 6-6.9 magnitude on Richter scale (combined).

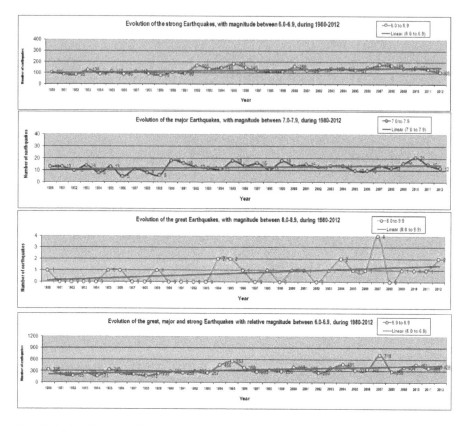

Figure 2. Earthquake trend evolution since 80's (blue thick line represents the increased linear trend and the coloured lines the frequency evolution for each type of earthquake category, strong, major or great):

4. Research course

The paper evaluate records of seismographs belonging to the international survey network over the last 30 years, assessing earthquakes frequency in order to detect evolution tendencies to be drawn. A simple linear correlation was used to categorize the trend of the seismic activity all over the world. Commonly the Earth seismic activity is almost constant in terms of frequency of earthquakes [3]. A possible increased tendency of earthquake activity was revealed studying the frequency of the principal earthquake types (such as: great, with the magnitude over 8, major with the magnitude of 7 - 7.9, and strong earthquake type of 6 - 6.9 magnitude on Richter scale), taking into consideration that an earthquake measuring 8 on the Richter scale is 10 times larger in term of ground motion than a 7 magnitude tremor, or 100 times larger than an earthquake measuring 6 magnitude, and so on. The results indicated an unusual increased seismic activity since the 90's, which is in contradiction with the generally constant trend of the previous decade. Based on lessons-learning approach, the activity of implementation of an earthquake resilient activity worldwide at local, regional or national level in the areas prone to earthquakes have to be assured by taking into account valuable recommendations of the risk managers involved into decisional planning, as indicated in the research paper.

5. Method used

Decision makers begin to understand that to save lives, they have to adopt an integrated, comprehensive and multi-hazard strategy for disaster risk reduction, regardless the type of the disaster management procedure. This strategy includes prevention, mitigation, preparedness, response, recovery and rehabilitation, therefore the following lessons learnt can be drawn:

5.1. Prevention measures

The latest tragedies highlighted the importance of the addressing of public buildings (such as: hospitals, schools, fire-fighter units, etc.) in the national earthquake protection policies;

A multi-hazard approach (earthquake plus tsunami) should be envisaged when response actions are planned. For example, access routes could have survived the earthquake but not the impact of the tsunami or some areas may remain flooded and therefore not able for rescue operations;

The constructions located in earthquake prone areas, erected before the last building regulation was put into force, have to be inspected in case of not complying with the norms, then have to be retrofitted or rebuild. A special attention should be done for retrofitting the construction for the most vulnerable socio-economical activities, which in case of earthquake could lead to severe loss of life, due to increased damages to the most vulnerable public areas (such as schools, fire-fighters units, hospitals, etc.), and interruption of public services (transportation, gas, electricity, water supply) by damaging the bridges, fall of power lines, pipelines rupture, etc;

The retrofitting works for all old buildings should take into account the new changing in the building resilience due to earthquake activity, taking into account the building codes for the specific earthquake area wherein the construction is located (for example, in Europe, the general rules for the assessment and strengthening of structures are available in the European Standard, Part 1-4 of Euro code 8, prEN 1998-3, and for other countries, the available guidelines in force). The designers and the constructors of the public units should pay more attention to structural issues;

As a result of the recent earthquakes, new building codes for earthquakes have to be introduced in the affected countries, including new seismic zoning of the whole country, with the purpose to improve the standards of building execution and maintenance. In addition, any dangerous structural changes implemented over the lifecycles of schools or other public buildings which can weaken the building strength have to be avoided. Therefore an increased activity of inspection should be undertaken regularly, according with the building code in force, in order to interdict any possibility for improvisation or structural changes, mainly for the public buildings. In areas prone to natural hazards, including earthquakes and tsunamis, it is necessary to constantly review and implement the proper building codes for constructions. In particular, the presence of adobe-built houses or improvised makeshift shelters can become disastrous;

In the coastal areas prone to tsunamis, it is necessary to implement prevention measures such as structural ones: tsunamis walls, sea walls, beach-long protection wall, automatic and manual closing water gates, evacuation routes and signing, establishing safer distances between different land use categories and the coastal line, depending to their economical activity, for minimizing the impact of possible tsunamis, or inexpensive protective lines of trees and dense vegetation, by planting local resistant trees species (for example mangroves in the tropical regions, coconut trees, etc.);

In addition, non-structural measures involve elaboration of tsunami vulnerability and risk maps, implementation of building codes and land use planning in order to define safe areas, education of the population regarding the behaviour in case of a tsunami wave, implementation of a seismic observation network system in relation to the possible detection of the tsunami generated by earthquakes, coupled with installation of alarming systems for the early warning of the population, studies for mapping the hazard vulnerability in the coastal areas characterized by intense socio-economical activity;

Early disaster events could be further analysed having a look at underwater sedimentary deposits in order to get a full picture of the vulnerability (including the case of marine deltas where new settled sediments once loosing stability can trigger tsunamis waves on the nearby coastal areas);

Although relatively reduced vulnerability of Stromboli type island (prone to underground landslides due to volcano material flow during eruptions) could be high due to holiday seekers and volcano tourists. Therefore the continuous activities of the volcano should seriously be watched and appropriate vulnerability analyses be performed. The focus should be put on landslides and/or lava flows due to volcanic activities; in addition, a multi-hazard approach could be useful as small earthquakes and/or tremors together with landslides trigger local tsunamis whose potential of destruction should not be underestimated. In the case of Stromboli one could promote structural actions (enforcing parts of the shoreline) and non-structural actions (educating the local population and especially instructing non-residents like tourists of potential signs of tsunamis). In the particular case of Stromboli volcano, which is of small size and not flat, it would be more efficient to manage an easy but effective concept of early-warning system (for ex., the use of loudspeakers, sirens) together with an evacuation system that allows moving the local population towards safer places in extremely short time;

The recurrence maximum time period, taken into consideration by nuclear engineers for a tremor in relation with a nuclear facility, that is the 10,000 years quake event, does not necessarily takes place after such a long period of time, and can occur anytime, even today or tomorrow, in the most earthquake prone areas all over the world, represented especially by the Pacific ring of fire, where the recent great earthquakes occurred;

The usual location for nuclear power plants are nearby large water available resources, sufficiently enough for assuring the cooling of water generated by the reactors, including tsunami prone areas nearby oceanic coastal shores. Consequently a higher location have to be selected for the backup power sources, and other electrical equipment for water pumps used to cool down the nuclear reactors following the automatic shut down due to largest possible tremor event ever recorded in the region, that means generally above 8 or 9 magnitude. Therefore every nuclear plant designs should take into account the resulting effects of this kind of event, including larger tsunamis than before experienced on a specific location chosen for nuclear development;

Periodical reevaluation of the nuclear power plant safety standards, depending of construction principle type e.g. light water cooled reactor (LWR), graphite-moderated, water-cooled reactor (RBMK), known as the Chernobyl type, heavy water moderated reactor (CANDU or AHWR), advanced gas cooled reactor (AGCR), liquid metal cooled reactor (LMF) or type of the nuclear fuel (uranium 253 and 258 or the most risky plutonium 239); NATECH scenarios (Natural Accidents that might trigger technical disasters) are to be considered, depending of natural hazards in the earthquake prone areas (e.g. landslides which may affect the land stability, storms or tsunamis which can flood the power generators, associated severe draught which may result in a water shortage in case of a water pipelines damage leading to nuclear fuel overheating), in order to avoid the worst case scenarios at a nuclear power plant, a nuclear leak due to melting down of the nuclear core, following failing of the cooling down of the exposed nuclear fuel rods.

5.2. Preparedness measures

The continuously monitoring of the areas prone to natural hazards, including earthquakes could lead to a better knowledge of the risk evolution of facing a possible disaster, also taking into account other vulnerability factors which can increase the probability of a disaster occurrence. Being known that many inhabited clusters could be closely located to an active tectonic area, and before some incipient earthquake activity will began, a detailed seismic analysis is necessary in order to detect the possible underground discontinuities. Generally speaking, even without having a historical evidence of earthquakes, worries can be raised regarding the overall seismic activity of a vulnerable area. In term of exposed population or industrial facilities, if an underneath fault is discovered, subsequent measures can be taken leading to a better preparedness activity for a possible earthquake;

The proper training of the personnel involved in emergency response and relief during natural disaster is essential for a better management of the emergency situations generated by an earthquake. Therefore constant simulation and drill exercises should be performed by the specialized personnel in order to be prepared in case of a major earthquake or for the possible forwarding aftershocks. An intense training program for the emergency personnel in the exposed areas should be performed using special trained sniff dogs and adequate equipment for increasing the preparedness capacity. Population should be also involved in the training drills, in order to become aware of the basic rules of survival and for recovery actions, to assure a better cooperation with the local authorities involved in the disaster mitigation activities;

The damage assessment scenarios for inhabited areas located in tsunami prone areas, on the coastal lines, will re-evaluate the mitigation capabilities in case of a real disaster and lead to a better response of the emergency services;

Countries located in tsunamis vulnerable areas should set their own national tsunami warning system, capable to watch and warn in due time the local inhabitants about any danger of producing a catastrophic event occurring nearby the inhabited area. For maintaining the awareness and the response capability of an already implemented tsunami warning system, simulation exercises should be periodically organized. Different responsibilities and tasks of the emergency personal involved in monitoring activities are reviewed, assuring the communication in real time of the emergency relief cruses about the probabilities of producing the disasters and assurance of warning the population;

The existence of the emergency stock of materials and means of interventions, located in the vicinity of the prone areas of natural hazards, including tsunamis, allows an optimized relief activity after a disaster in the region, assuring a successful intervention activity and minimization of loss of lives and damages to the properties. It is crucial to have sufficient stock (including tents, blankets, medicine) available in order to support people that have fled from the tsunami;

An efficient preparedness measure depends of timely early warnings issued by the authorities following an earthquake with high magnitude, which often constitute the triggering factor for the tsunami;

Area that had been affected by similar events in the past should create a disaster prevention platform; it could help in better identifying vulnerable areas and/or weaknesses in preparedness activities;

Evacuation routes should be generated on the basis of flood maps and availability of shelters. If no natural shelters (hills, mounds, berms) are available it is advisable to construct vertical shelters.

It should be clear that living in houses which are built 1 - 3m above sea levels, a high level of preparedness is required in the case a tsunami hit;

Already established safety zones, implemented in the planning of the coastal areas, will lower the risk of the highly vulnerable areas, both by earthquakes tremors and tsunami waves, therefore a multi-hazard approach in emergency planning would be advantageous. Preceding disasters, like a heavy earthquake, could (partly) destroy evacuation routes and assembly places; therefore a multi-hazard approach (earthquake plus tsunami) should put particular emphasis on having such routes and places secured. Moreover, the emergency planning should take into account that subsequent disasters or inconveniences may happen and request alteration of early plans, i.e. heavy rainfalls which, in turn, produce landslides and mudflows.

Subsequently, people in emergency shelters had again to be redistributed in (different) safe locations;

In the particular case of Stromboli type volcanic island, due to the continuous activities of the volcano, constant preparedness is absolutely required, that is availability of responsible persons issuing the alarms, instruction non-residents, keeping free the evacuation routes;

On small islands telecommunication back-up system should be kept operating in order to start rescue operations;

The nuclear facilities located in the earthquake prone areas should have drilled in advance holes for vent up hydrogen released from the water cooling down reactor. The holes should be positioned at the top of the main building covering the nuclear reactor and containment vessel. This means preventing the hydrogen build up and risk of deflagration which might cause radioactive emissions, in case of core overheating due to breakdown of the cooling system. These hydrogen releases due to radiolysis may take place also because of the nuclear rods exposure in case of lowering down the water level in the cooling water pools with nuclear depleted material found inside the main buildings of the nuclear power plant;

Every nuclear power plant should take into consideration the availability of a pool of human resources to be used as a supplementary intervention in catastrophic event. In addition, a clean-up facility building located a few kilometers away from the main reactor facilities, including shelters large enough to host the emergency shifts for extended intervention in case of a nuclear incident. Such an action is recommended when the number of the normal available working shift personnel can not assure a proper emergency intervention in case of power failure and reestablishing the cooling down capabilities of a possible crippled nuclear reactor due to the twin action of a large scale tremor and subsequent tsunami event.

5.3. Response measures

The endowment of the rescue teams with special equipments and means of intervention in case of emergency situations is essential for an efficient response, increasing the chance for saving lives and reducing the economical impact of the natural disasters, including earthquakes. In the aftermath of the disaster, many persons can be rescued beneath the rubble thanks to the sniffer dogs and hi-tech ultrasound equipment both from the national level or foreign emergency teams;

The existence of the communication routes through all remote communities within a prone area for natural disaster, including tsunami, is an essential factor for undertaken an efficient response activity in case of a disaster event;

For minimizing the pressure of the local community in case of disasters, the existence of an insurance system for the houses and goods against the natural disasters, including earthquakes is very efficient. This is due to the indemnity of the affected people, automatically covered by the insurance companies. The financial coverage of the response action will not be affected, in case of producing some damages. Commonly, in the aftermath of an earthquake, the only compensation of the homeless people in the affected areas are the subvention from the state and foreign aid organizations, in order to assure the economical income for a normal social life. Anyway it couldn't cover always integrally the loss, in the absence of a national-wide efficient insurance system;

In the hazard prone areas where a certain disaster is present, the recovery activities are difficult to undertake, for example in arid regions there is the possibility that water tubes are broken triggering major damages. Response teams must be ready to get water lines repaired in short time;

In the rehabilitation phase the focus should be put on economical recovery and social sustainability within the affected communities. Therefore long-term intervention development programs have to be set up in the affected areas, for the benefit of the most vulnerable communities, mainly focusing on income generating projects;

The multi-hazard feature of the inhabited areas and population vulnerability, as a result of the economical developing, could worsen the condition of the affected population in case of a natural disaster, superposing the effect of more hazards. A prime task of the international assistance in the affected regions is the strengthening of the capacity to respond to future disasters in the area, because some regions could have been already suffering from the effects of other hazards before the earthquake, or to withstand to the associated hazards of the main event (such as aftershocks, tsunamis, fires due to broken gas pipelines or from the damaged reservoirs of the affected boats or cars carried by the waves into the houses walls, liquefaction and landslides, mudflows, etc.);

A prompt response activity in case of a natural disaster, including tsunami, is related to the existence of an already implemented, "Plan of emergency and intervention", at the level of local and central public authorities. It clearly stipulates the competencies and the activities during each phase of the emergency intervention for rehabilitation and clearance of the disaster effect. The plan should be constantly revised in order to assure the updating of the information with the changes in land planning activities at the level of the community, or modifications intervened in the structure of the emergency staff personal in charge with the response activities;

Rescue operators have to count with a lot of destruction and uninhabitable houses thus having to maintain a huge number of refugees over a long period;

The response capability in coastal areas, in the case of a tsunami event, should rely on the effectiveness of the early warning system for tsunami, which allows an efficient preparedness measure. In some vulnerable coastal areas the travel time for tsunami to reach the coastal area is very short (for example the Mediterranean region), generally in less than 10 min after start, due to relatively shallow and low step offshore bottom morphology. Consequently the period of time until the tsunami alert is initiated should be very short, in relation to an existing efficient alarm capability of the population and the emergency relief crews;

Automatic unmanned (anti-radiation proof for humans) crane coupled with long range powerful water pumps near a water source for spraying at distance large volume of waters, should be available for all nuclear facilities located in the earthquake prone areas, including tsunamis. These special intervention equipments, including remote surveying robots with dosimeters, should be used in the event of a nuclear cooling down operation failure, following larger tsunamis that might drawdown the back up pumps used for emergency intervention. In addition, a longer enough power cable to be switched on at an existing nuclear facility from an outside existing power source, generally a mile longer, should be available to connect by emergency the main nuclear unit of reactors in case of power failure due to earthquake tremor or subsequent tsunamis. Large barge should be available nearby for transporting freshwater in case of a nuclear accident at a plant located at the sea shore, in order to cool down the reactors, because the marine salt water damages irrevocably the nuclear facility.

6. Information to the public

In the areas prone to natural disasters, including earthquakes, at the level of the regional or local administration, hazard vulnerability and risk maps should be available for all decisional factors involved in the management of this type of disaster but also for dissemination to the general public in order to be informed about the dangers nearby the inhabited areas;

The proper information of the population from the vulnerable areas to the earthquakes about the risk reduction issues and the possibility to reduce the vulnerability of their houses by applying correct building codes, is highly necessary. The using of the new building materials, such as iron or iron coated concrete beams, together with the traditional ones such as clay bricks, without respect to any elementary building code, sometimes worsened the strength of a construction, and put an increased risk of the inhabitants. For example, the use of the iron beam for strengthening and to allow the extra-store constructions, together with traditional materials (clay bricks), could increase the vulnerability in case of a possible earthquake, as well as in the case of recently affected areas by earthquakes, where multi-store buildings collapsed and produced more casualties than in a possible destruction of a one store house;

It is necessary to create a knowledge platform to disseminate information at the local level, to educate people about the risk reduction issues in case of an earthquake. For example, the existence of some water and food supplies, also some vital medicines in case of chronic diseases, available in case of trapping inside a house can increase the life expectancy in case of earthquake, which could produce the collapse of the inhabited house;

The adequate information regarding the situation nearby an affected area by a recent earthquake lead to a more donor support from the surrounding communities and countries. An information booklet and a Website describing the earthquake effects during the relief operations can bring more donor support and can contribute together with the information press for a humanitarian appeal from the international community;

The ongoing information of the public regarding the actions to avoid a tsunami wave (such as: the clear indication of the escape routes, the avoidance of the exposed coastal areas during the tsunami, urgent deployment to higher places, etc.) will lead to an adequate behaviour of the population in case of a real disaster, limiting the number of affected individuals;

The information of the public about subsequent effects of a technological disaster (oil terminals and refineries, mostly located in the tsunamis prone coastal areas) or natural hazards in travel or inhabited area, including tsunamis, and about the presence of other possible accompanied triggered disasters, following an earthquake, such as landslides or rock falls, by all available means (police agents, local broadcasting, tv news, papers, warning panels, etc.), lead to avoid the risk and limits the consequences in the aftermath of a natural disaster;

The case of Stromboli type islands, visited by numerous foreign tourists, requires permanent, effective and multi-lingual instruction of residents and non-residents, i.e. leaflets let to those arriving, pictograms let in hotel rooms, warning signs put on beaches and nearby paths;

Populations should be kept informed by local authorities on the possible restriction zone, generally following an accident at a nuclear reactor due to the impact of a twin event of tremor and the subsequent tsunami wave. The restriction zone is declared generally as an exclusion zone for population, excepting the nuclear plant emergency personnel and fire fighter units, and is particularly coffined at a specific distance radius to the crippled nuclear reactor, commonly of value of tens of miles around the radiation source;

Radiation self-detection equipment (dosimeters) for personal use should be available for the population individuals located nearby nuclear facilities, or the persons travelling nearby, for auto monitoring of the radiation doses (e.g. the hourly radiation dose is 0.1 micro Sievers - $\mu Sv/hour$). In case of exceeding the normal dose, depending on instructions from the emergency supervising personnel, a decontamination procedure is required (e.g. shower with water and soap washing); Food (milk and fresh harvested vegetables) and water nearby a crippled nuclear facility can be immediately affected by a nuclear leakage due to a catastrophic failure of the cooling down the nuclear reactor, or incidents at the nuclear rods being exposed, due to the wind dispersion (e.g. as far as 100 km radius far to the radioactive source);

Main radioactive isotopes (e.g. Iodine 131, Xenon-133, Krypton-85 and Caesium 137), produced during a nuclear accident due to subsequent tsunami of an earthquake event, can immediately affect the health on long term, due to the carcinogenic effect. Special medication for radiation prevention should be used only on the certified medical surveillance, because the main antidote, for Iodine 131, the iodine salts (e.g. potassium iodine) is available just for a time window of 4 days, when the results are affective (e.g. for avoiding the accumulation in thyroid gland by aerial way), and the self medication with other similar inhibitors (for example iodine salt), can shift the effective period before the radioactive cloud is atmospherically drifting on a certain vulnerable inhabited zone.

7. Status

Contemporaneous seismic activity as well as complementary volcanicity are genetically linked to Cenozoic plate kinematics, involving interacting plates and/or intra-plate rifting steaming from triple junctions. Upper mantle heterogenic seismic structures are intimately related to plate breaking and motion.

A series of natural facts are to be taken into consideration in order to approach the causes of such unusual trend of increasing major earthquake frequency after 1990 which led to destructive earthquakes in "classic" areas but also in areas not specifically known as prone area.

A cause of increasing trend of seismic activity may be induced by internal factors related to global tectonics. It is marked by intracrustal-subcrustal structural, sedimentologic and magmatic processes creating shallow or deep areas for large magnitude earthquakes., e.g. coupling convergence rate, age of subduction, lithosphere type, trench sediment thickness and so on.

Reality or mere coincidence, concurrent supracrustal processes at global scale may affect the Earth's structure and related sensitive tectono-seismic spots. Of them the global warming is considered by a large part of the academic world as major process with implications at atmospheric, hydrospheric, biospheric and lithospheric levels that represents the so-called Critical Zone of the Earth. So far Cenozoic eco-climate change was taken into consideration in order to explain seismic differences of orogenic regions based on sediment thickness, i.e. effect of coupling between tectonic and erosion.

The need for detailed analyses of the effects of the global warming and the assessment of all the aspects of the environmental factors, are due to the necessity to control the natural hazards at the level of the planet Earth, involving the approach of a global analysis. Therefore, to study the interrelation between global warming and the earthquakes can be made just analysing the involved phenomena (earthquakes, global warming, tectonic evolution) at the world scale level, taking into account all relevant aspects of the involved hazards, making reference to the historical evidence and data records.

Useful information can be provided by conclusions of the experts involved in the analyses of the core samples from the ice drillings 3000 m deep in Greenland, performed in early '90s. The unusual enriched content of sulphate found in the ice cores at a certain depth proved an episode of unusual intense volcanic activity, which took place at 7000 BC, induced by the tectonic instability due to the rapid defrosting of the continental ice sheet, because of a warmer climatic episode. The paleo-environmental reconstruction of the last major volcano activity occurred on earth, at 7000 BC, was a result of the analyses conducted on the ice drilling samples from glaciers by a research program performed in Greenland, through the European Science Foundation [6].

8. Results

The analyses of the earthquakes frequency trend all over the globe, in the recent years, correlated with the actual tendency of defrosting the ice from the polar regions [7], allow the study of presumable recurrences in future, of a similar event of volcano increasing activity and subsequently tectonic disturbance, as a result of the defrosting evolution of the actual glaciers due to global warming. The possible correlation between the analysed earthquake data and the actual ice sheet evolution, which covers actually 10% of the total crust (where the continental area covered by the ice is reducing due to the increased global temperature), can induce serious consequences over the tectonic stability of the earth, respectively the frequency and magnitude of the related phenomena, such as subsequent volcanoes activity which can be induced by the plates movements and evolution, and generating earthquakes.

According to specific environmental evaluations that claim that the actual trend of global warming is continuing, in the next hundreds of years the continental ice will disappear. The rapid defrosting of the continental ice could lead also to some secondary tectonic effects due to the release of the equivalent pressure of the ice load. If the assumption of the paleo-climatologists is correct, a similar phenomenon, an increased warming episode, like the actual global trend, could lead to an unexpected increased tectonic activity, with unpredicted impact over the humans and surrounding environment. Therefore the actual increased trend of the earthquakes frequency could be a global indicator of the tectonic stress due to rapid defrosting of the continental ice sheet.

9. Further research

Furthermore the general lessons have to be implemented urgently by the risk managers involved in the activities of updating and implementing the building codes, seismic risk zoning and regulation, in order to avoid in the future any other misjudges of the earthquakes hazard, for minimizing the loss of human lives and material damages.

10. Conclusions

Analysing the possible increased tendency of earthquake activity (Table no. 1), in order to clarify the cause of the unusual increased trend of the earthquakes frequency in certain periods of times after the 90's, a common fact was that all these recently past events surprised the local population as well as local and national level risk managers, because the hit areas were not considered before specific historically earthquake prone zone, so the building codes were not updated for a real seismic zone (including major cities as Kobe or Islamabad). The paradoxical issue of increased trend of earthquakes just after '90 was never been tackled seriously before. Generally it is considered that just 10% of the total energy from tectonic plates movement are transformed in earthquakes, and remain 90% converted in other forms of energy due to rock displacement and heating up processes [5]. A constant increasing trend of the Earth's earthquake energy, revealed by our analysis over the last 30 years seismic records worldwide, could indicate a shifting of the remain 90% of the tectonic energy, normally dissipated in plates interactions, towards earthquakes. For the the first time in modern history, were recorded in the same day two great earthquakes more than magnitude 8 on Richter Scale, in the same area (off the west coast of Northern Sumatra, during 2012), instead of a smaller aftershock of the same tremor, usually not exceeding a lower range magnitude (the aftershock shouldn't exceed a magnitude 7 on the Richter scale). That's mean we will witness a future increasing in the earthquake pattern trend, which may have profound implications at a global scale, in our understanding of Earth dynamics.

Magnitude/year	1980	1981	1982	1983	1984	1985	1986	1987	1988	1989	1990
8.0 to 8.9	1	0	0	0	0	1	1	0	0	1	0
7.0 to 7.9	13	13	10	14	8	13	5	11	8	6	18
6.0 to 6.9	105	90	85	126	91	110	89	112	93	79	109
	1991	**1992**	**1993**	**1994**	**1995**	**1996**	**1997**	**1998**	**1999**	**2000**	
	0	0	0	2	2	1	0	1	0	1	
	16	13	12	11	18	14	16	11	18	14	
	96	166	137	146	183	149	120	117	116	158	
2001	**2002**	**2003**	**2004**	**2005**	**2006**	**2007**	**2008**	**2009**	**2010**	**2011**	**2012**
1	0	1	2	1	1	4	0	1	1	1	2
15	13	14	14	10	10	14	12	16	21	15	12
126	130	140	141	148	148	178	168	144	151	134	108

Table 1. Evolution of the Earthquakes frequency (no/magnitude/year) during 1980-2012

All around the globe, in the earthquake prone areas, scientists monitor carefully earthquake activity, because many agglomeration centres, including large areas like Tokyo and Bucharest (later the capital of Romania, considered the most prone capital with a similar earthquake activity as Mexico City, in the opinion of the most celebre seismologist, Charles F. Richter) are expecting a devastating event, according to the statistics (Tokyo is expecting "the big one" earthquake following the last major event in 1923, so called "Kanto earthquake", and in Romania the same Vrancea source earthquake, with the last major event in 1977, with more than 2 billion US dollars in damage and 1500 fatalities).

Another explanation is that, following the global climatic changes, a large part of ice Polls sheet started melting (unprecedently during the summer of 2012, for the first time the Greenland ice sheet was partially melted at the surface, far exceeding with 100 years the climatologists previsions), so large volume of water were released into the ocean triggering potential changes in the global plate tectonic equilibrium. Taking into account that Antarctica (Southern Pole continent) is covered with snow and ice of almost 2000 m height, equivalent in weight of a real continent, whose melting can destabilise the established continental plates equilibrium. These sudden melting (which in terms of geological ages has never been experienced so fast until now in the whole Earth's geological history), might influence the global earthquake trend, a possible precursor of changes in the pattern of global plate tectonic movement. What is however certain is the fact that earthquakes are geological hazards of endogenous origin, and what is uncertain is the global warming itself and the potential influence of exogenous factors over crustal/sub-crustal settings. Consequently, discerning mere speculation from evidence is still a priority.

The needs for increasing the resilience of the communities all over the world lead to more detailed studies on both small and large scale in order to try to explain the connection among factors which interact naturally on the Earth. The lessons learning activity based on the analysis of the recent tremors data all over the world can improve the preventive, preparedness and intervention means of the earthquake vulnerable areas.

Author details

Septimius Mara[1*] and Serban-Nicolae Vlad[2]

*Address all correspondence to: maraseptimius@yahoo.com

1 Ministry of Environment and Forests, Romania, Bucharest, Romania

2 Faculty of Ecology and Environmental Protection, The Ecological University, Romania

References

[1] Airinei, Ş. (1972). Geophysics (in Romanian), Ministry of Education and Science, Bucharest University

[2] Georgescu, E. S. Bucureştiul şi seismele (in Romanian), Editura Fundatiei culturale Libra, Bucuresti, (2007).

[3] Lazarescu, V. (1980). Physical Geology (in Romanian), Technical publisher, Bucharest

[4] Peter, M. Shearer, "Introduction To Seismology", Cambridge University Press, 1999http://earthquake.usgs.gov/hazards/about/workshops/thailand/downloads/ CSMpp1_History.pdf,accessed 24 March (2012).

[5] U.S. Geological Survey, National Earthquake Information Center- Earthquake list http://earthquake.usgs.gov/earthquakes/eqarchives/year/eqstats.php, updated on 2012 (accessed 03 March 2012)

[6] Science daily: Underwater Earthquakes Geophysicists Discover Slippery Secret Of Weaker Underwater Earthquakes (2007)http://www.sciencedaily.com/videos/2007/ underwater_earthquakes.htm,accessed 15 April (2012).

[7] Zielinski, G. A, Mayewski, P. A, Meeker, L. D, Whitlow, S, Twickler, M. S, Morrison, M, Meese, D. A, Gow, A. J, & Alley, R. B. US Global Change Research Information Office-Increased Volcanism Linked To Climatic Cooling During The Period From 5000 To 7000 B.C. reference: Record of Volcanism Since 7000 B.C. from the GISP2 Greenland Ice

Core and Implications for the Volcano-Climate System, Science, *http://www.science-mag.org/content/264/5161/948.short)*,accessed 2 May (2012). , 264, 948-952.

[8] National snow and ice data center (NSIDC) Sea Ice Decline Intensifies- (report on 28 September 2005); http://weathertrends.blogspot.com/2005/09/sea-ice-decline-intensi-fies.html,accessed 15 June (2012).

Chapter 4

Seismotectonic and the Hypothetical Strike – Slip Tectonic Boundary of Central Costa Rica

Mario Fernandez Arce

Additional information is available at the end of the chapter

1. Introduction

The studied area is comprised of the Central Volcanic Range (CVR) of Costa Rica, the northwest flank of the Talamanca Cordillera, and the space between them, known as the Central Valley of Costa Rica (Figure 1). The Central Valley separates volcanic rocks of the CVR from intrusive rocks of the Talamanca Cordillera. The zone is characterized by low seismicity in the north and high seismicity in the South (Montero, 1979; Montero & Dewey, 1982; Montero and Morales, 1984).

Astorga et al. (1989, 1991) proposed the existence of a strike-slip fault across Costa Rica extending from the Pacific to the Caribbean and passing through the central part of the country. Fan et al. (1993) stated that a diffuse transcurrent fault zone trending northeast-southwest and composed of various subparallel strike-slip faults exists in Central Costa Rica. According to Fan et al. (1993), the fault zone extends from the Pacific coast to the Caribbean across central Costa Rica, and may represent a possible plate boundary for the proposed Panama Block. Jacob et al. (1991), Fisher et al. (1994) and Marshall (2000) assured that the strike-slip tectonic boundary traverses the Central Valley of Costa Rica. The prior proposals were mentioned in many other works [Seyfried et al. (1991), Fisher y Gardner (1991), Güendel y Pacheco (1992), Fan et al. (1992), Goes et al. (1993), Lundgren et al. (1993), Marshall et al. (1993), Gardner et al. (1993), Escalante y Astorga (1994), Protti y Schwartz (1994), Montero (1994), Marshall (1994), Montero et al. (1994), Fernández et al. (1994), Barboza et al. (1995), Marshall y Anderson (1995), Marshall et al. (1995), Suárez et al. (1995), Di Marco et al. (1995), Colombo et al. (1997), Güendel y Protti (1998), López (1999), Lundgren et al. (1999), Montero (1999), Yao et al. (1999), Quintero y Güendel (2000), Montero (2001), Trenkamp et al. (2002), Husen et al. (2003), Linkimer (2003), Montero (2003), DeShon et al. (2003), Norabuena et al. (2004), Pacheco et al. (2006), Marshall et al (2006), Camacho et al. (2010)] what spread the idea of the existence of a tectonic boundary in Central Costa Rica.

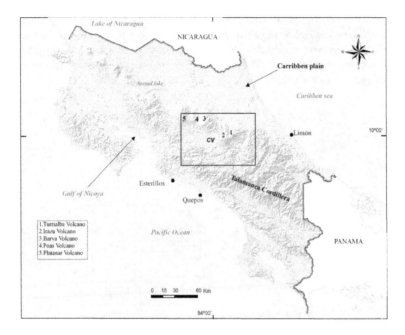

Figure 1. The area of interest is indicated by the rectangle and covers part of the Central Volcanic Range (numbers mark key volcanoes), the Central Valley (CV) of Costa Rica and Talamanca Cordillera. The Central Valley is a depression located between the ranges and contains the largest population centers of Costa Rica.

Older references have been used to support the hypothetical tectonic boundary of Central Costa Rica [Van Andel et al. (1971), Stoiber y Carr (1973), Burbach et al. (1984), Adamek et al. (1988), Carr y Stoiber (1990) and Mann et al. (1990)] but they are not appropriate to justify the boundary because they refer to a segmentation in the Cocos Plate not in the Caribbean Plate.

This paper analyses and discusses the seismicity and faulting of Central Costa Rica in search for evidence of the strike-slip fault proposed by Astorga et al (1989, 1991), the subparallel strike-slip fault system reported by Fan et al. (1993) and the plate boundary trace in the Central Valley of Costa Rica suggested by Jacob et al. (1991), Fisher et al. (1994) and Marshall et al. (2000).

2. Data and method

Available data on faulting, historic earthquakes, instrumentally recorded shocks and source mechanisms are provided in this work. Information on faulting is compiled from Fernández & Montero (2002); and Denyer et al., (2003). The seismic data has come from the data file compiled by the RED SISMOLOGICA NACIONAL (RSN: ICE-UCR) operated by the University of Costa Rica (UCR) and the Instituto Costarricense de Electricidad (ICE). This seismic network monitors the seismic activity of Costa Rica with 20 analog, short-period vertical-component seismometers (black triangles, Figure2) and 9 digital three-component stations (open triangles, Figure 2). The signals from analog stations are telemetered to the University of Costa Rica at San Jose where they are digitized by an A/D converter and recorded on a PC computer running the SEILOG data acquisition program. The station spacing is densest in the study area and in western Costa Rica.

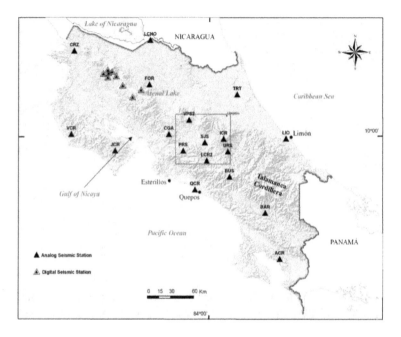

Figure 2. Seismic stations of the Red Sismologica Nacional (RSN: ICE_UCR) shown with triangles. Black triangles are analog stations. The digital stations are indicated by open triangles.

Historical data on earthquakes are from Rojas (1993). The recent seismicity includes shallow earthquakes of depth equal to or smaller than 30 km and intermediate/deep earthquakes with depths larger than 30 km. Both data subsets span from 1992 through 2009 and were extracted from databases of 4845 (shallow) and 7756 (intermediate/deep) events. The range of duration magnitudes is 1.8-6.2 and the average is 2.8.

The subset of 865 high-quality shallow events includes 382 located by Fernández (1995) and 82 more by Fernández (2009). They were located with 5 or more stations (7 average) and 2 readings of S wave. Their average rms residuals and horizontal and vertical errors in location are 0.3 sec, 1.8 and 2.0 km respectively. The average azimuthal gap between stations used in the hypocenter determinations is 149.2° and the average distance to the closest station is 15.3 km.

The subset of intermediate/deep earthquakes includes only those locations showing vertical error smaller than 10 km.The average latitudinal and longitudinal component of the location

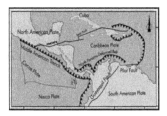

Figure 3. Tectonic Setting. Costa Rica is located on the western extreme of the Caribbean Plate. The border between this plate and the Cocos plate is the Middle American Trench (MAT) located off the Costa Rican Pacific coast. Other tectonic boundaries are the Polochic-Motagua-Chamalecon Fault System (PMCHFS), the Panama Fracture Zone (PFZ) and the North Panama Deformed Belt (NPDB). From Fernandez et al. (2004)

errors for this kind of events are 6.35 and 6.2 km. Their average rms residual is 0.4 sec and the average distance to the closest station is 30.6 km.

Earthquakes were located using P and S wave arrival times and the SEISAN program (Havskov and Ottemøller, 1999) which includes a version of the Hypocenter. A 1-D seismic velocity structure, determined by seismic refraction in northern Costa Rica (Matumoto et al., 1977), is used by the RSN to locate earthquakes in Costa Rica. Fernández (1995) located earthquakes of Central Costa Rica with the 3-D velocity structure of Protti (1994). Fernandez (1995) and Protti et al.(1996) found no significant differencies between earthquake locations obtained with both the 1-D and the 3-D models.

Focal mechanisms for major events in the area were determined by using the first motion of P-waves. The P-wave first motion data were plotted on an equal area projection of the lower hemisphere. The search of fault planes was restricted to events with at least 9 reported first motions. These inversions were performed with the FOCMEC program (Snoke et al., 1984).

3. Tectonic setting and geology

Central America is an active island arc built up by the northeast subduction of Cocos lithosphere beneath Caribbean plate. The junction of these plates forms the Middle American Trench (MAT), the western boundary of the Caribbean plate (Figure 3). The present convergence rate increases along the trench from about 7.3 cm/yr off Mexico and Guatemala to 8.5 cm/yr in western Costa Rica (DeMets 2001). Seismicity suggests that the northeast dipping slab has descended to a maximum depth of 200 km in western Costa Rica (Protti et al., 1994) and to only 70 km off southern Costa Rica. (Arroyo, 2001). The subduction became shallower at the southern terminus of MAT in response to a buoyant submarine ridge (Cocos Ridge) that arrived to the trench ~5 Ma (de Boer et al., 1995), causing a decrease in the volcanic activity. The subduction of the Cocos ridge, which rises almost 2 km above the surrounding seafloor, generates high uplift and significant deformation of the whole arc in front of the present subducting ridge. A major geologic effect produced by the subduction of Cocos plate in southern Costa Rica is the uplift of the Talamanca Cordillera.

The Talamanca Cordillera is a Miocene plutonic-hypabissal volcanic complex that extends by 180 km from central Costa Rica to western Panama. Major Tertiary volcanic complexes are present in this range but large and young strato-volcanic complexes are absent, a consequence of the significant elevation of the range (de Boer et al., 1995) and the shallow, high-angle subduction in southern Costa Rica [60° according to Arroyo (2001)]. This range is the highest topographic feature of Central America and, therefore, of the Caribbean plate. This elevation is possibly related to the subduction of Cocos Ridge (Kolarsky et al., 1995).

The Central Volcanic Range is a chain of andesitic stratovolcanoes trending northwest, parallel to the MAT. The CVR Consists of five massifs-Platanar, Poas, Barva, Irazú, Turrialba--and several pyroclastic cones associated to the main volcanoes. This cordillera covers an area of 5150 m^2 and its maximum topographic feature is Irazu volcano (elevation 3400 m). The volcanic activity at the present-day edifices commenced in the Late Cenozoic and has continued throughout the range until the present. The current activity consists of fumarolic emissions and hot intra-crater lakes. Barva and Platanar are dorman volcanoes of this range.

The Central Valley is a narrow trough (15 km wide, 70 km long) between the Central Volcanic and Talamanca ranges. Late Tertiary and Quaternary volcanic rocks, believed to be part of the current volcanic edifices forming the Central Volcanic Cordillera, are present in this valley as well as some Miocene sedimentary sequences.

4. Faulting

Previous works, field investigations and assessments of neotectonic features via airphotos indicate that deformation of central Costa Rica occurs in three geographical areas: the Central Volcanic Range, the Central Valley and the northern flank of the Talamanca Cordillera (Figure 4).

The Central Volcanic Range faulting is divided into three sub-zones: Irazu Volcano, Bajo de la Hondura, and Poas Volcano. Irazu is a zone of northwest-trending, short length (< 20 Km), normal faults and some northeast faults whose displacement is also normal (Figure 4). Within the Bajo de la Hondura zone, in the low between Irazu and Barba volcanoes, are the south-north trending Hondura and Patria normal faults and the strike-slip Lara fault. At Poas, in the northwest extreme of the Central Volcanic Range, the southeast- northwest-striking Viejo, Carbonera and Angel faults border the volcano.

Over decades Costa Rican geologists have considered faulting absent in the Central Valley of Costa Rica. Geologic maps show several faults in the borders of the valley but only few within it (MIEM, 1982, MINAE, 1997, Tournon & Alvarado, 1997, Denyer et al., 1993). Such faults probably exist but are difficult to recognize because of the volcanic and concrete surface cover. Among the better known faults of this area are the Alajuela and Escazu. Alajuela is a 28-km long east-west reverse fault and Escazú seems to have reverse and strike-slip component (Fernández and Montero, 2002). In the last decade additional high-quality seismic data have begun to illuminate important structures within the valley. Fernandez and Montero (2002) mapped three more faults in the valley (Cipreses and Río Azul). An interesting finding is the

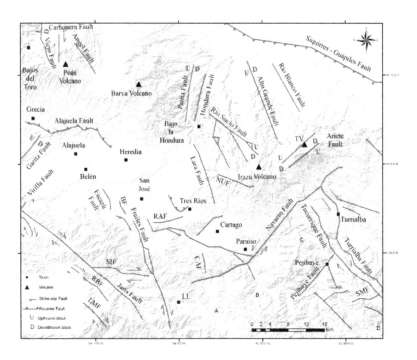

Figure 4. Faults mapped in Central Costa Rica. Triangles mark volcanoes; squares show cities or towns. The aligned volcanoes mark the longitudinal axis of the Central Volcanic Range. The cities of Cartago, San José, Heredia and Ala-juela are located in the Central Valley of Costa Rica. The Escazú and Aguacaliente (AF) faults define the southern boundary of the Central Valley. Faults located southeast of the Navarro fault belong to the Talamanca Cordillera. PF: Picagres Fault, BF: Belohorizonte Fault, RAF: Río Azul Fault, SIF: San Ignacio Fault, RBF: Resbalon Fault, LMF: La Mesa Fault, CIF: Cipreses Fault, NUF: Nubes Fault, CAF: Cangreja Fault, SMF: Simari Fault, ATF: Atirro Fault, PCF: Pacuare Fault, LL: La Lucha, SP: Santiago de Puriscal, U/D: normal faults howing relative motion: U, upthrown block; D, down-thrown block. Sawteeth along solid lines indicate thrust fault. Strike-slip arrows represent strike-slip faults.

extension of the Aguacaliente and probably Rio Azul faults under the surface of San Jose, the capital of Costa Rica, which represent a significant hazard for that city.

At the Talamanca Cordillera faults trend predominantly northwest with varying fault lengths and slip directions. The most important faults in this area are Atirro, Navarro, Aguaca-liente, Frailes-Escazu, and Jaris. In the east, the dextral Atirro Fault is the major structure, and it splits into two branches (the Tucurrique and Turrialba faults). At the northern rim of the range, the Aguacaliente Fault marks the boundary between the range and the Central Valley. Trench excavations across the Navarro, Aguacaliente, and Orosi faults have been conducted in order to date the most recent ruptures and to identify periods of dormancy (Woodward-Clyde, 1993). Soil development along faulted surfaces and scarp morphometry was used to determine the relative deformation rates across the segments. At the Navarro fault, the trench shows evidence of faulting within the unconsolidated sediment section, where sediment deformation features are present. These features include lineaments such as small strike-slip and reverse faults, along fault line locations mapped during field studies. Results suggest that faulting has occurred during the Holocene, but movement is likely disseminated over a broad zone (100 m) instead of being concentrated along any single fault plane. At Aguacaliente, one trench intersected a trace that offset the soil horizon by approx- imately 30-35 cm (Woodward-Clyde, 1993). The apparent displacement was normal and a dated carbonizad log suggested that the last movement on this fault occurred less than 3700 years ago. On a trench across the Orosi fault in Cartago, Costa Rica, the most significant finding was a set of fractures cutting all the soil units and suggesting normal dip slip, down to the east. The fractures coincide with the steepened facet of the break in slope on the colluvial fan (Woodward-Clyde, 1993).

The NW-striking Frailes-Belohorizonte-Escazu fault zone extends 30 km. The fault zone is marked by scarps, slope changes, and offsets of aligned stream channels and divides. Accord- ing to Fernandez and Montero (2002) this fault system combines dextral and uplift movement and consists of discontinuous fault traces.

The Guapiles-Siquirres fault runs along the base of the Central Volcanic Range, and therefore, marks the boundary between that range and the Caribbean plain. It is a combination of two continuous reverse faults, Guapiles in the North and Siquirres-Matina in the South (Denyer et al., 2003). Soulas (1989) proposed that the Siquirres-Matina fault is the prolongation of the North Panama Deformed Belt within the territory of Costa Rica. The Guapiles-Siquirres fault is characterized by high topographic relief with uplifted terraces and deep-narrow river valleys over much of its length (Soulas, 1989). Linkimer (2003) extends this large fault to Aguas Zarcas de San Carlos (not shown) for a total distance of 150 km.

Neither the strike-slip fault proposed by Astorga et al. (1989) nor the set of subparalel strike- slip faults suggested by Fan et al. (1993) were found in the studied area. The trace of the strike-slip tectonic boundary suggested by Jacob et al. (1991), Fisher et al. (1994) and Marshall et al. (2000) neither was found within the Central Valley of Costa Rica. The most impor- tant east-west faults, the faults required by the hypothetical strike-slip tectonic boundary, of the Central Valley are Aguacaliente and Alajuela. The first one shows a component of normal slip and the second is a tipical reverse fault that connects with the Garita fault whose slip is normal.

5. Seismicity

5.1. Historical seismicity

Well-documented historical earthquakes data from 1700 to 2006 have been analyzed in this work to understand the seismicity of central Costa Rica. Our catalog contains 15 events (Table 1), 7 of which occurred in the Poas Volcano seismic zone, one near Irazu volcano, one west of the city of Heredia and 6 south of the Central Valley. Figure 5 shows a well-defined cluster at the western end of the Central Volcanic Range (Poás volcano area) and another at the northern flank of the Talamanca Range (south of the Central Valley).

No.	Name	Latitude	Longitude	Year	Magnitude	Seismic Zone
1	Barva earthquake	10.1000	-84.2000	1772	5.6	Poás
2	Cartago earthquake	09.8250	-83.9300	1834	5.2	South of Central Valley
3	Alajuela earthquake	09.9500	-84.2670	1835	5.8	Puriscal
4	Cartago earthquake	09.8416	-83.9100	1841	5.8	South of Central Valley
5	Alajuelita earthquake	09.8300	-84.1000	1842	5.4	South of Central Valley
6	Fraijanes earthquake	10.1380	-84.1840	1851	5.5	Poás
7	Fraijanes earthquake	10.1380	-84.1830	1888	5.8	Poás
8	Tablazo earthquake	09.8166	-84.0333	1910	5.2	South of Central Valley
9	Cartago earthquake	09.8416	-83.9100	1910	6.4	South of Central Valley
10	Toro Amarillo earthquake	10.2333	-84.3000	1911	6.1	Poás
11	Sarchí earthquake	10.1916	-84.2750	1912	6.2	Poás
12	Tres Ríos earthquake	09.8666	-84.0000	1912	5.2	South of Central Valley
13	Paraíso earthquake	09.8083	-83.8800	1951	5.2	South of Central Valley
14	Patillos earthquake	10.0250	-83.9083	1952	5.5	Irazú
15	Toro Amarillo earthquake	10.2333	-84.3166	1955	5.8	Poás

Table 1. Historical earthquakes in Central Costa Rica (Rojas, 1993)

The historical seismic data correlate well with previously identified faulting. For instance, at the Poas seismic zone 5 earthquakes are located along the northwest-trending faults that border the volcano from south to west (Figure 5). It is quite probable that the Carbonera and Viejo faults were responsible for the Bajos del Toro (1911, 1955) and Sarchi (1912) earthquakes. The damage zones described for the Fraijanes earthquakes (6 and 7 on Figure 5) suggest that the source could be the Angel fault. To the southeast, the epicenters of historical earthquakes are located on the periphery of the Talamanca Cordillera, where most form an alignment along the Aguacaliente fault (the Cartago earthquakes of 1834, 1841 and 1910 and the Tres Rios earthquake of 1912). The 6.4 Ms Cartago earthquake (1910) and the 5.2 Ms Tres Rios earthquake (1912) appear to be in the same seismogenic context; the 1910 event possibly strained the northwest segment of the Aguacaliente fault and, two years later the accumulated strain was released originating the Tres Rios earthquake. A similar situation could have happened at Poas when Sarchi earthquake followed the 1911 Bajos del Toro earthquake.

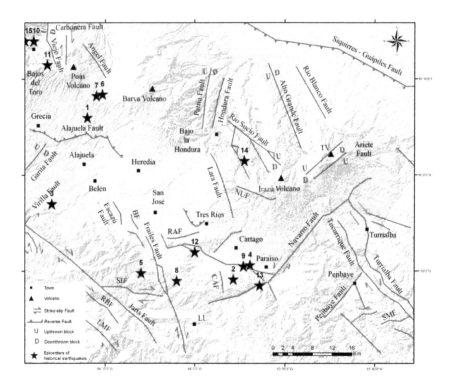

Figure 5. Map showing the historical earthquakes in Central Costa Rica. Stars mark the epicenters of historical earthquakes of the last two centuries. The number near each star is the number of the event in Table 1.Earthquakes 4 and 9 and 6 and 7 share the same epicentral area, respectively. Triangles indicate volcanoes. AF: Aguacaliente Fault, PF: Picagres Fault, BF: Belohorizonte Fault, RAF: Río Azul Fault, SIF: San Ignacio Fault, RBF: Resbalon Fault, LMF: La Mesa Fault, CIF: Cipreses Fault, NUF: Nubes Fault, CAF: Cangreja Fault, SMF: Simari Fault, ATF: Atirro Fault, PCF: Pacuare Fault, LL: La Lucha, SP: Santiago de Puriscal.

Additional strong evidence for the correlation between historical earthquakes and faulting comes from isoseismal maps. Montero & Morales (1988) found elongated intensity contours that clearly surround the known source of these events. For the Cartago, Tres Rios and Fraijanes earthquakes, the contoured intensity distributions relate the earthquakes to northwest-trending faults, suggesting that the Angel and Aguacaliente faults participated in the generation of those events. Bajos del Toro, Sarchi and Patillos events have northeast-trending damage areas that disagree with the fault orientation; in these cases the lack of reports northward the source could affect the geometry of the isoseismal map.

This historical seismicity is considered upper-crustal seismicity by White (1991) and White & Harlow (1993). The later authors pointed out that upper-crustal earthquakes are spatially distributed along the volcanic front of the whole of Central America; they appropriately called them volcanic-front earthquakes and stated that these earthquakes pose the greatest hazards for the population.

A final remark about this seismicty deals with its connection with large Costa Rican earthquakes. Upper-crustal destructive earthquakes of central Costa Rica in the last one hundred years coincided with large earthquakes that took place in the country. In 1904, a 7.2 Ms magnitude subduction earthquake happened in southern Costa Rica and also 6.8 Ms event southwest of the Central Valley, and five years later the Cartago (1910), Tablazo (1910), Bajos del Toro (1911) and Sarchi (1912) earthquakes occurred in Central Costa Rica. Similarly, in 1950 the largest earthquake reported in Costa Rica occurred, a 7.7 Ms magnitude subduction event that was followed by the Paraiso (1951), Patillos (1952) and Bajos del Toro (1955) earthquakes. These data suggest that destructive events of central Costa Rica may represent seismicity triggered by large subduction events.

All of this evidence suggests that historical earthquakes did not occur randomly, and moreover, they did not form any lineament in an east-west direction that supports the existence of a tectonic boundary with that orientation in central Costa Rica. Those events are clearly associated with faults that have been recently mapped.

5.2. Instrumental seismicity

The epicentral distribution of 865 shallow earthquakes (0-30 km) recorded by RSN during the period 1992-2009 is plotted in Figure 6. This shallow seismicity is not uniformly distributed over the study area, that is, there are seismic clusters separated by zones of low level seismicity. On a rough scale, the seismicity of Talamanca is higher than the seismicity of the Central Volcanic Range. In the Central Valley the seismicity has the lowest rate for the whole area.

The volcán Irazu is a zone of seismic swarms that resemble volcano/tectonic. According to Fernández et al. (1998) there have been seismic swarms at Irazu in 1982, 1991, and 2007. The pattern of these swarms is a large number of very small earthquakes with few moderate events of magnitude 4 or so, but no clear mainshock larger than the other events. They have occurred on short fault of the zone, especificaly on Elia, Ariete and Nubes.

At the Bajo de la Hondura, a trough between the Irazu and Barva volcanoes, scarce but permanent seismicity has been recognized. It is a seismicity of magnitude smaller than 5. One of the recent major events was the magnitude 4.4 earthquake that occurred there on August 21,1990, at 13 km depth. The main sources of this activity are the Hondura, Patria and Lara faults.

The seismic activity at Poas is mainly composedd of swarms and sporadic strong earthquakes. The swarm activity consists, like the Irazu activity, of a hundred of small earthquakes generated during one or two months. Fernandez et al. (in prep.) have recognized seismic swarms at Poas in 1980, 1990 and 1999 According to their location, the last swarms at this area was generated by Carbonera and Angel faults. A strong 6.2 Mw magnitude earthquake hit the zone on January 8, 2009 killing 25 people and destroying many houses, several bridges and the route to Cinchona. In adition, the earthquake triggered many landslides in the epicentral area. As a consequence of such earthquake the village of Cinchona (Figure 7) had to be reubicated. The economic losses from the destructive earthquake are estimated in $492 million (Laurent,

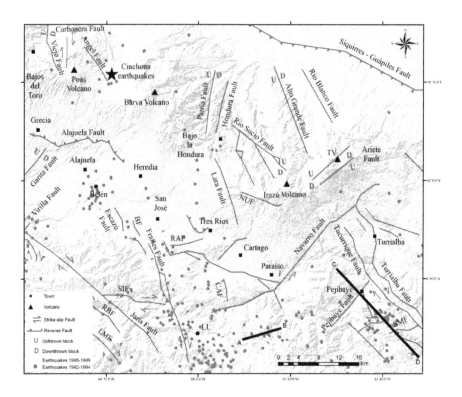

Figure 6. Shallow (0-30 km) seismicity of Central Costa Rica from 1992 through 2009. Several clusters represent the most important seismic zones in the studied area. Crosses are seismic event for the 1995-2009 period. Diamonds are earthquakes located by Fernandez (1995) and black circles represent earthquakes located by Fernandez (2009). Lines A-B and C-D indicate traces of cross sections. AF: Aguacaliente Fault, PF: Picagres Fault, BF: Belohorizonte Fault, RAF: Río Azul Fault, SIF: San Ignacio Fault, RBF: Resbalon Fault, LMF: La Mesa Fault, CIF: Cipreses Fault, NUF: Nubes Fault, CAF: Cangreja Fault, SMF: Simari Fault, ATF: Atirro Fault, PCF: Pacuare Fault, LL: La Lucha, SP: Santiago de Puriscal.

2009). The event was located in the eastern flank of the volcano at 4 km deep and was generated by the Angel fault.

In the Talamanca Cordillera the seismicity is spread all over the area but there are also dense clusters at Pejibaye, south of Cartago and Santiago de Puriscal (Figure 6). Two of these clusters correspond with isolated seismic sequences (Pejibaye and Puriscal) and the other one with a zone of permanent seismicity (La Lucha).

The Pejibaye July 10 1993 (Mc = 5.3) earthquake, together with the Mc = 4.9 July 8 foreshock two days before and the Mc = 4.8 aftershock three days later represent the most extensive and well-recorded seismic sequence in the eastern part of central Costa Rica (Fernandez, 2009). These earthquakes and many aftershocks occurred within a small area of northwest and northeast-trending faults. The event's depths are relatively shallow and can be associated with Simari fault which, according to focal mechanisms, is strike/slip with a high normal component.

Figure 7. The village of Cinchona after the 2009 Cinchona Earthquake. The earthquake changed the geography of the area. Courtesy of Joanna Mendez.

Puriscal was a quiet seismic zone before 1990 but in that year there began one of the highest concentrations of seismic activity of Costa Rica in recent decades. This activity was triggered by a large earthquake from the Pacific Coast. Thousands of micro earthquakes were generated in Puriscal in the December 1990-June 1991 period, almost 30 events of Mc > 4.0 and the main event of Mc = 5.7, the Piedras Negras earthquake.

La Lucha is the most seismically active zone in central Costa Rica, however a large percentage of its present-day seismicty is microearthquake activity (Mc < 3.0). Although the epicentral distribution is diffuse, a northwest trend can be recognized, and this trend is in good agreement with that of the Frailes Fault. The main structural features associated with La Lucha seismicity are Frailes and Navarro faults.

While the Central Volcanic and Talamanca Ranges have significant seismicity (Fernandez, 1995; Fernandez et al., 1998) the number of recorded earthquakes and their magnitudes reflect very little activity within the Central Valley. During more than 20 year of records, the background microseismicity of this valley is represented as scattered low-level activity (Fernandez, 1995). The best known and well- defined concentration of earthquakes in the valley is in Belen and seems to be associated with the Escazu fault. A more recent manifestation of seismicity has been observed in the metropolitan area of San Jose in the last 5 years; it consists of $2 <$ Mc < 4 earthquakes whose epicenters appear to define a NW-striking lineation that coincides with the northwest end of the Rio Azul and Aguacaliente faults. In the southern border of the central valley there are seismic sources with relatively high rates of seismicity such as the Escazu and Aserri faults, both related to the Frailes-Belohorizonte Escazu fault system. The Aguacaliente fault, responsible for the 1841 and 1910 Cartago earthquakes, has had little activity in the last three decades.

In an effort to see if earthquakes define faults, seismicity cross-sections were carried out in the studied area. Due the low number of earthquakes in some cases and the nearness between faults in other cases only two seismic cross sections were calculated, one at the Pejibaye seismic zone and other eastward of La Lucha. In the cross section A-B (Figure 8a) the hypocenters seem to define a inclined plane that dips 75° northeast, which suggests that a high-angle fault is the responsible for this seismic activity. The cross section C-D (Figure 8b) reveals that the dense seismicity cluster along the Pejibaye seismic zone is generated by an almost vertical fault. Fernandez (2009) reported a fault dipping 76° northwest as the cause of this seismicity.

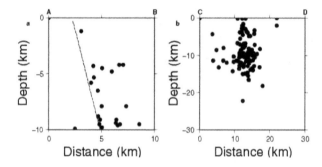

Figure 8. Seismic cross sections A-B and C-D outlined in Figure 6.

5.3. The seismic anomaly of Central Costa Rica

Recent earthquake epicenters from 1992 through 2009 were plotted on a map of Costa Rica in order to show the characteristic local pattern of seismicity that is possibly associated with tectonic features. The plot displays a wide zone of high subduction and crustal seismic activity in Central Costa Rica which coincides with a diffuse zone rather than with a narrow longitudinal area (Fernández et al., 2007). The seismicity forms an anomalous big cluster composed of smaller clusters (Figure 9) but despite the considerable concentration of earthquakes, epicenters of either the big or smaller clusters fail to delineate any large and single NE or EW fault plane.

To know whether or not the seismicity pattern is related to a hypothetical strike-slip tectonic boundary, we examined the depths of the earthquake clusters. We would expect shallow seismogenic source locations for a strike-slip tectonic boundary but deep (greater than 30 km)

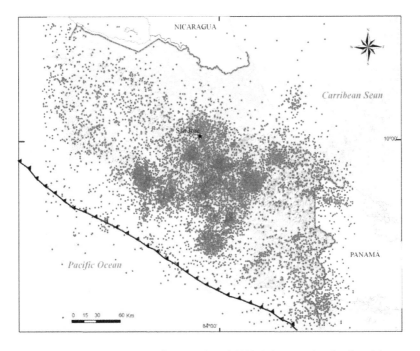

Figure 9. Background seismicity in Costa Rica from 1992 through 2009. Circles are earthquakes. Several clusters represent the most important seismic zones of Costa Rica. The sum of these clusters generates a zone of concentrated seismicity in Central Costa Rica. MAT: Middle American Trench.

source locations for subduction zone earthquakes. Because 80% of the present-day seismicity of Central Costa Rica is shallow, we expect earthquake concentrations to be above a subduction decollement.

To test whether the seismic origin is in the subduction zone or from a much shallower transform fault earthquakes with depths in the range of 30–90 km were plotted at intervals of 10 km (Figure 10). Costa Rican earthquakes are distributed over all depths with deeper clusters to the northeast. The cluster in figure 8a approximately coincide with the results of DeShon et al. (2003) who found that earthquakes occur above 30 km depth, 95 km from the trench offshore Central Costa Rica. Our results suggest a source for the anomaly related to the subduction process, perhaps subducted seamounts on the Cocos plate that generate larger stress fields than nearby smooth subducted areas of the same plate, causing the high intraplate and interplate seismicity in central Costa Rica. Bilek et al. (2003) stated that shallow, smaller-magnitude seismicity is more common in regions of seamounts subduction than in the smoother region subducting off northern Costa Rica, suggesting that subduction of topographic highs localizes seismicity. Von Huene et al. (2004) indicate that subducted seamounts appear to remain attached to the underthrust plate more than 100 km landward of the trench axis as indicated by clustered earthquakes beneath the shelf and local uplift along the coast.

This is in excellent agreement with our results, which support the seamount domain of Central Costa Rica as the cause of the seismic anomaly.

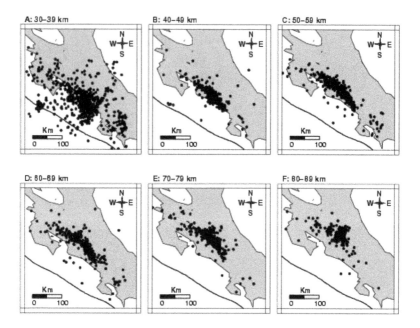

Figure 10. Cumulative numbers of located earthquakes, separated into six depth ranges, within or near the Costa Rican territory. These graphs plot earthquakes detected and located by the Red Sismologica Nacional (RSN: ICE-UCR) from 1992 to 2009. Depth ranges are in km.

Figure 11 shows a set of seamounts on the Cocos plate between the Fisher mounts and the Quepos plateau. The seamounts form a subducting rough zone that collides with the Caribbean plate generating stress, deformation and weakening of the continental crust. Onshore, in front of this zone is the seismic anomaly of Central Costa Rica. The ocean bottom in the Cocos plate between Quepos plateau and Cocos Ridge is almost flat and the seismic level in front of this rectangular area is relatively low (Figures 11). These facts also suggest that sea mounts play an important role in generating seismicity in Costa Rica. They apparently increase intraplate and interplate earthquakes onshore and therefore, in absence of them the seismic activity in Central Costa Rica would probably be lower than the current activity.

6. Focal mechanisms

P-wave first motion is used to determine focal mechanism solutions. However, first-motion observations will frequently be in the wrong quadrant because of incorrect first-motion

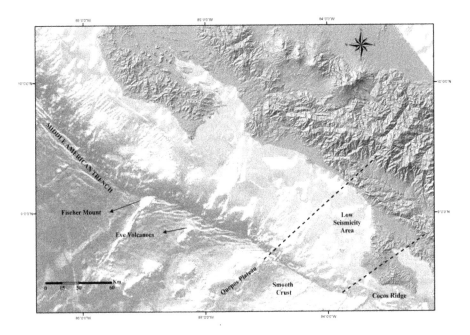

Figure 11. The Cocos-Caribbean tectonic boundary in front of the Costa Rican Pacific coast is the Middle American Trench. Large seamounts (Fisher Mount, Eve volcanoes, Quepos plateau) are being subducted under the Caribbean plate just in Central Costa Rica. This process causes high stress and seismicity. From Ranero and von Huene, 2000.

direction, inappropriate earthquake velocity model, station polarity reversals and incorrect direct P-arrival picks due to low signal-to-noise ratios. The method requires enough data to ideally determine fault-plane solutions. Few data or incorrect first motion observations may generate more than one or many focal mechanism solutions and changes in the earthquake location or in the seismic velocity model can significantly affect the distribution of observations on the focal sphere, changing the best-fitting focal mechanism solution. Low magnitude earthquake and seismometers locates near the nodal planes between the compressional and dilatational quadrants of an earthquake do not produce strong first motions which made difficult to determine focal mechanisms.

Because the studied area is characterized by microseismicity and truly few intermediate-magnitude earthquakes, it is really difficult to obtain a large number of reliable focal-plane solutions in central Costa Rica. After a strict selection of seismic events of the last 18 years, we only found 16 reliable focal mechanisms (Table 2, Figures 12 and 13). They show considerable variation in the sense of motion which probably reflects movement on preexisting planes of weakness that are geometrically favorable for slip but not necessarily aligned with a plane of maximum shear stress. The events exhibit reverse, normal and strike-slip faulting.

Number	Date	Latitude	Longitude	Mag	Depth	RMS	EH	EZ	AZ	Dip	Rake
1	90/12/22	09.883	-84.334	5.7	14.6	0.28	0.7	1.7	252.5	63.0	30.7
2	92/11/02	09.887	-83.766	3.4	06.2	0.21	0.4	0.7	060.3	72.8	-58.4
3	92/11/03	09.921	-84.138	4.1	06.5	0.30	0.6	0.9	269.0	40.0	58.0
4	92/11/12	09.745	-84.013	3.5	16.8	0.27	0.7	2.1	097.6	51.1	145.6
5	93/01/20	09.979	-84.183	3.7	11.6	0.35	0.8	2.0	230.0	90.0	45.0
6	93/05/07	09.705	-83.767	3.7	03.8	0.28	0.6	0.8	236.9	56.4	-10.3
7	93/07/09	09.756	-83.615	4.3	12.6	0.30	3.2	4.8	239.6	68.5	-57.5
8	93/07/10	09.776	-83.686	5.3	12.8	0.31	2.2	3.2	262.37	75.9	-32.4
9	93/07/13	9.735	-83.615	4.9	12.4	0.22	3.0	2.9	240.5	43.9	-22.2
10	93/07/14	09.701	-83.809	3.9	06.7	0.59	0.8	2.0	224.9	45.9	-76.0
11	94/01/11	09.812	-84.142	3.5	16.8	0.21	0.6	1.1	110.4	65.4	79.0
12	94/10/29	09.867	-84.064	3.3	06.6	0.30	0.7	0.5	253.0	84.0	-40.0
13	96/05/23	09.850	-83.988	3.1	11.4	0.36	0.7	2.0	097.0	74.0	-53.0
14	96/05/26	10.090	-83.660	4.0	14.9	0.40	2.2	3.6	210.0	50.0	-90.0
15	99/07/18	10.206	-84.228	3.2	04.8	0.31	1.0	0.5	359.0	66.1	-26.3
16	09/01/08	10.194	-84.177	6.2	03.6	0.60	2.6	2.6	025.0	47.5	-37.0

Note. Mag.: Magnitude, RMS: root-mean-square, EH: horizontal error, EZ: vertical error, AZ: Azimuth.

Table 2. Parameters of focal mechanisms.

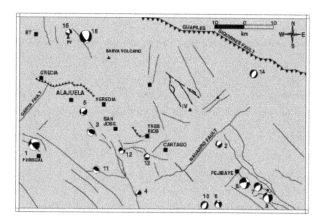

Figure 12. Faulting and focal mechanisms. Small lettered stereo projections are fault-plane solutions for 16 carefully selected earthquakes. BT: Bajos del Toro, PV: Poas Volcano, IV: Irazu Volcano.

FAULT PLANE SOLUTIONS

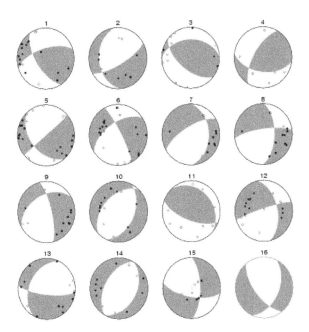

Figure 13. P-wave first motion focal mechanisms, determined using pspolar routine of GMT (Graphic Mapping Tools). In all cases more than 9 P-wave polarities were used. Open circles represent downward first motions, black circles represent upward first motion.

Focal mechanisms near Pejibaye (6, 7, 8, 9, and 10) show nearly normal-slip along planes striking northeast, suggesting a possible association with a northeast-trending faults. At Puriscal, the fault-plane solution (1) is strike-slip with reverse component. That solution indicates right-lateral motion along the northeast striking nodal plane. Based on the destruction near Alajuela associated to the correspondent earthquake Montero (2001) chose that plane as the fault plane and proposed the Virilla fault as the responsible for the earthquake. However, the strike of the selected nodal plane is close to the orientation of the Picagres fault.

Fault-plane solutions for events from Frailes-Escazú faults (3, 4, 5, 11, 12) show thrust and strike-slip motion with a strong reverse component (3, 4, 11). These solutions suggest northwest striking faulting, in good agreement with the strike of the mapped faults. Event 13 suggests a high normal component along the Aguacaliente fault. When resolvable, the focal mechanisms of small to moderate sized earthquakes (M< 4.5) in the Poas area show predominantly strike-slip motion (15, 16). The fault-plane solution for the 2009 Cinchona earthquake (16) is oblique with high normal component (Rojas et al., 2009).

Another important limitation to obtain more and better focal mechanisms in Central Costa Rica is the instrumentation used to detect them. We are still using short period, one component seismic sensors to detect and locate the seismicity. Due to this, the resolution of the strike for the occurring mechanisms depends on the readings at only few stations in many cases. In the future it would be more appropriate to compute the focal mechanisms using waveform inversion (Dreger & Helmberger, 1993; Zhu & Helmberger, 1996; Herrmann et al., 2008; D´ Amico et al., 2010; D´Amico et al., 2011).

7. Discussion

The faulting, high seismicity and strike-slip focal mechanisms do not define a consistent east-west shear zone in central Costa Rica. Strike-slip deformation in central Costa Rica is inter-preted as a result of the elastic strain accumulation in the upper plate due to the subduction of seamount domain and Cocos Ridge under the Caribbean Plate. The fault orientation may reflect the northeast movement of Cocos plate, stresses caused by the subduction of sea mounts, and the compression of the Cocos Ridge in southern Costa Rica, where the rate of convergence between Cocos and Caribbean plates is maximum (DeMets, 2001). This high rate and the south-north sliding of the Cocos plate along the Panamá Fracture Zone could be creating a favorable environment to form northwest lateral tears (as Frailes, for instance).

White & Harlow (1993) studied the destructive shallow earthquakes in Central America and found a concentrated seismicity in the volcanic front. According to them, this volcanic front is a zone of dextral strike-slip driven by oblique subduction. Large earthquakes as that of Managua in 1972 and Tilarán in northern Costa Rica in 1973 were strike-slip earthquakes. These data indicate that strike-slip motion within the Caribbean Plate is not concentrated in Costa Rica but it is present all over Central America (Quintero & Guendell, 2000)

Fan et al. (1993) proposed that left-lateral strike-slip motion in central Costa Rica occurs on various sub-parallel strike-slip faults that comprise a diffuse northeast-southwest strike-slip fault zone. This is inconsistent with Astorga et al. (1989, 1991) who proposed an east-west trend for the fault system of Central Costa Rica. But the proposal of Astorga et al (1989, 1991) is not supported by the data described here.

Fischer et al. (1994) stated that the seismicity after Cóbano (1990) and Limón (1991) earthquakes are constrained in a diffuse zone of faulting oriented west-east along the Central Valley of Costa Rica and that the variety of faults may reflect an early stage of a developing shear zone. In this work all currently mapped faults and lineaments are included and we find the same faulting pattern that Arias & Denyer (1991) attribute to a north-south compression that affects Costa Rica since late Miocene-Pliocene. The distribution of earthquakes and focal mechanisms indicate that seismic activity occurs on both northeast and northwest trending faults. There-fore, the seismicity mentioned by Fisher et al. (1994) is not likely to be due to incipient faulting but to preexistent faulting reactivated by the collision of Cocos Ridge with the Caribbean Plate (Denyer & Arias, 1991) and by faults reactivated after large earthquakes.

Strike-slip deformation along plate boundaries is often distributed among several parallel faults (Brink et al, 1996) and shear zones are overprinted by numerous foliation-parallel brittle faults (Cunningham, 1996). Offset strike-slip faults may be connected by intervening pull apart basins but this geometric pattern is not well defined in Central Costa Rica. There are parallel faults but they do not follow a preferential direction and not all of the parallel faults are strike-slip in type. Observing the fault distribution and orientation near the Central Valley of Costa Rica, we see parallelism between the most important: Alajuela, Aguacaliente and Frailes-Escazu faults (northwest extreme). But the Alajuela Fault is a very well-known reverse fault and the Frailes-Escazu also seems to have a strong reverse component according to Denyer et al. (1993), Fernández & Montero (2002) and our results in this work. Focal mechanisms and an excaved trench suggest that in contrast the Aguacaliente fault has a significant normal component. If this is so, the central Valley of Costa Rica would not be a pull apart basin unless it represents a developed strike-slip fault system where strike-slip faults have gradually evolved into oblique thrusts or thrusts (Fuh et al., 1997).

Marshall et al. (2000) attributed the deformation of Central Costa Rica to the subduction of Cocos Ridge and the seamount domain and proposed an E-W deformation front that propagates northward into the overriding volcanic arc, as the tectonic boundary between the Caribbean plate and the Panama block. But even this deformed belt requires a set of EW strike-slip faults along its northern edge, located in the Central Valley of Costa Rica. However, the EW strike-slip faults, and therefore the EW strike-slip motion, are absent in the studied area and most active faults of that area are northwest. DeMets (2001) and Norabuena et al. (2004) estimated trench-paralell motion of the Costa Rican forearc to northwest at a rate of 7 and 8 mm/yr respectively. They suggest interseismic and post-seismic effects from forearc faults and the subduction interface, diffuse extension at the trailing edge of the forearc sliver, partitioning of slip between multiple forearc faults, northwest striking right-lateral strike-slip faults and vertical axis rotation of smaller blocks defined by short, northeast striking, left-lateral "book-shelf" faults as the multiple cause of the observed motion. In the same way, northeast motion could have multiple explanations.

Von Huene et al., (2003) assure that subducted seamounts are causing deformation and weakened of the upper plate which steepness the slope above them, generating great potential for tsunamigenic landslides. The sea mounts destroy the frontal prism and uplifts the continental crust. Since this result it is clear that subducted seamount play an important role in the deforming the upper plate in central Costa Rica.

8. Conclusion

There is a seamount domain off central Costa Rica and intense crustal deformation and high seismicity onshore, in front of this seamount domain. The deformation includes an x-pattern faulting in which both northeast and northwest faults are active and have high seismicity. Focal mechanisms of small-magnitude earthquakes show normal, reverse and strike-slip motion along some faults of the studied area. Most of the historical earthquakes, the largest earthquakes of the zone, suggest northwest motion along the Viejo, Carbonera, Angel, Frailes and Aguacaliente faults.

The strike-slip fault of Costa Rica proposed by Astorga et al (1989) and the set of subparalel strike-slip faults suggested by Fan et al. (1993) were not found in the studied area. Neither the trace of the hypothetical strike-slip tectonic boundary, which according to Jacob et al. (1991), Fisher et al. (1994) and Marshall et al. (2000) cut the Central Valley of Costa Rica, was not found in that valley.

According to our data, there is no a clear and well defined east-west strike-slip fault system in Central Costa Rica that might represent a tectonic boundary. The anomalous deformation and seismicity of central Costa Rica is more related to the subduction of sea mounts than to the proposed hypothetical strike-slip tectonic boundary for Central Costa Rica.

9. Data and resources section

- Earthquake data were provided by the Red Sismologica Nacional (RSN) operated by the Costa Rican Electricity Company and the University of Costa Rica. They cannot be released to the public.

- Some plots were made using the Generic Mapping Tools version 4.2.1 (www.soest.hawaii.edu/gmt; Wessel and Smith, 1998).

Acknowledgements

Thank to personnel of both Central America Seismological Center (CASC) and the Red Sismológica Nacional (RSN: ICE-UCR) for providing data to carry out this investigation. My gratitude to Sara Kruse for comments and suggestions that greatly improved the manuscript. Also thanks to Cindy Solis and Jonnathan Reyes for their help in processing the data and preparing the figures. The author is grateful to CONICIT for financial support through FORINVES program.

Author details

Mario Fernandez Arce*

Address all correspondence to: mario.fernandezarce@ucr.ac.cr

Escuela de Geología, Universidad de Costa Rica, Programa PREVENTEC, Red Sismológica Nacional (RSN: ICE-UCR). San José, Costa Rica, Central America

References

[1] Adamek, S, Frohlich, C, & Pennington, D. Seismicity of the Caribbean-Nazca boundary: Constraints on microplate tectonics of the Panama region. J. Geophys. Res., (1988). , 93, 2053-2075.

[2] Arias, O, & Denyer, P. Estructura geológica de la región comprendida en las hojas topográfica Abra, Caraigres, Candelaria y Río Grande, Costa Rica. Rev. geol. Amér. Central, (1991). , 12, 61-74.

[3] Arroyo, I, Sismicidad y neotectónica en la región de influencia del proyecto Boruca: hacia una mejor definición sismogénica del Sureste de Costa Rica. 162 pp. Tesis de Licenciatura, Escuela de Geología, Universidad de Costa Rica, (2001).

[4] Astorga, A, Fernández, J, Barboza, G, Campos, L, Obando, J, Aguilar, A, & Obando, L. Cuencas sedimentarias de Costa Rica: Evolución Cretácico Superior-Cenozoica y potencial de Hidrocarburos.-Symposium on the Energy and Mineral Potential of the Central American- Caribbean Region, San José, Costa Rica, March 6-9, 1989, Circumpacific Council: 23 , 1989.

[5] Astorga, A, Fernández, J, Barboza, G, Campos, L, Obando, J, Aguilar, A, & Obando, L. Cuencas sedimentarias de Costa Rica: Evolución geodinámica y potencial de hidrocarburos. Rev. Geol. Amer. Central, (1991). , 43, 25-59.

[6] Barboza, G, Barrientos, J, & Astorga, A. Tectonic evolution and sequence stratigraphy of the central Pacific margin of Costa Rica. Rev. Geol. Amer. Central, 18, (1995). , 43-63.

[7] Bilek, S, Schwartz, S, & Deshon, H. Control of seafloor roughness on earthquake rupture behavior. Geology, (2003). , 31(5)

[8] Brink, U, Katzman, R, & Jian, L. Three-dimensional models of deformation near strike-slip faults, J. Geophy Res., (1996). , 101(B7)

[9] Burbach, G, Frohlich, C, Pennington, W, & Matumoto, T. Seismicity and tectonics of the subducted Cocos plate. J. Geophys. Res., (1984). , 89, 7719-7735.

[10] Camacho, E, Hutton, W, & Pacheco, J. A New at Evidence for a Wadatti-Benioff Zone and Active Convergence at the North Panama Deformed Belt, Bull. Seism. Soc America, N. 1, (2010). , 100, 343-348.

[11] Carr, M, & Stoiber, R. Volcanism, in The Caribbean region, The Geology of North America, vol., H, edited by G. Dengo, and J. Case, Geological Society of America, Boulder, Colorado, (1990). , 375-391.

[12] Colombo, D, Cimini, G, & De Franco, R. Three-dimensional velocity structure of the upper mantle beneath Costa Rica from a teleseismic tomography study. Geophys. J. Int., (1997). , 131, 189-208

[13] Cunningham, W, Windley, B, Dorjnamjaa, D, Badamgarov, J, & Saandar, M. Late Cenozoic transpression in southwestern Mongolia and the Gobi Altai-Tien Shan connection, Earth and Planetary Science Letters, (1996). , 140, 67-81.

[14] Amico, D, Orecchio, S, Presti, B, Gervasi, D, Guerra, A, Neri, I, Zhu, G, & Herrmann, L. R. B., Testing the stability of moment tensor solutions for small and moderate earthquakes in the Calabrian-Peloritan arc region. Boll. Geo. Teor. Appl., doi:10.4430/bgta0009,(2011). , 52, 283-298.

[15] Amico, D, Orecchio, S, Presti, B, Zhu, D, Herrmann, L, & Neri, R. B. G., Broadband waveform inversion of moderate earthquakes in the Messina straits, Southern Italy, Physics of Earth and Planetary Interiors, doi:j.pepi.2010.01.012, (2010). , 179, 97-106.

[16] De Boer, J. Z, Drummond, M. S, Bordelon, M. J, Defant, M. J, Bellon, H, & Maury, R. C. Cenozoic magmatic phases of the Costa Rican island arc (Cordillera de Talamanca), in Mann, P., ed., Geological Society of America Special Paper, Geologic and Tectonic Development of the Caribbean Plate Boundary in Southern Central America, (1995). (295), 35-55.

[17] Demets, C. A new estimate for present-day Cocos-Caribbean plate motion: Implications for slip along the Central American volcanic arc, Geophys. Res. Lett, (2001). , 28

[18] Denyer, P, & Arias, O. Estratigrafía de la región central de Costa Rica, Rev. Geol. Amer. Central, (1991). , 12, 1-59.

[19] Denyer, P, Arias, O, Soto, G, Obando, L, & Salazar, G. Mapa Geologico de la Gran Area Metropolitana, (1993).

[20] Denyer, P, Montero, W, & Alvarado, G. Atlas Tectonico de Costa Rica, Editorial Universidad de Costa Rica, (2003). , 81.

[21] Deshon, H, Schwart, S, Bilek, S, Dorman, L, Gonzalez, V, Protti, M, Flueh, E, & Dixon, T. Seimogenic zone structure of the Middle America Trench, Costa Rica, J. Geophys. Res. 108 (B10), 2491, (2003).

[22] Marco, G, Baunmgartner, P., Channel, J., Late Cretaceous-early Tertiary paleomagnetic data and a revised tectonostratigraphic subdivision of Costa Rica and western Panama, in Mann, P., ed., Geologic and Tectonic Development of the Caribbean Plate Boundary in Southern Central America: Boulder, Colorado, Geological Society of America Special Paper 295, (1995).

[23] Dreger, D. S, & Helmberger, D. V. Determination of source parameters at regional distances with single station or sparse network data. J. Geophys Res., (1993). , 98, 1162-1179.

[24] Escalante, G, & Astorga, A. Geología del Este de Costa Rica y el Norte de Panamá. Rev. Geol. Amér. Central, v. esp. Terremoto de Limón: (1994). , 1-14.

[25] Fan, G, Beck, S, & Wallace, T. A Diffuse Transcurrent Boundary Boundary in Central Costa Rica: Evidence From a Portable Aftershock Study (Abstract), Eos. Trans., AGU, 73, 345, (1992).

[26] Fan, G, Beck, S, & Wallace, T. The Seismic Source Parameters of the 1991 Costa Rica Aftershock Sequence: Evidence for a Transcurrent Plate Boundary. J. Geoph Res. 98, B9: 15,759-15,778, (1993).

[27] Fernández, J, Botazzi, G, Barboza, G, & Astorga, A. Tectónica y estratigrafía de la Cuenca Limón Sur. Rev. Geol. Amér. Central, v. esp. Terremoto de Limón: (1994)., 15-28.

[28] Fernández, M. Análisis sísmico en la parte central de Costa Rica y evaluación del hipotético sistema de falla transcurrente de Costa Rica, Tesis de maestría, Universidad Nacional Autónoma de México (UNAM), 85 , 1995.

[29] Fernandez, M, Camacho, E, Molina, E, Marroquin, G, & Strauch, W. Seismicity and neotectonic of Central America, in: Bundschuh, J., Alvarado, G. (eds), Central America- Geology, Resource and Hazards; Taylor & Francis Customerr Services, Andover, United Kingdom, 1340 , 2007.

[30] Fernandez, M, Escobar, D, & Redondo, C. Seismograph Networks and seismic observation in El Salvador and Central America, Geological Society of America Special Paper (2004). , 375, 257-267.

[31] Fernandez, M, & Montero, W. Fallamiento y Sismicidad del Area entre Cartago y San José, Valle Central de Costa Rica, Rev. Geol. Amer. Central, (2002). , 26, 25-37.

[32] Fernández, M, Mora, M, & Barquero, R. Los procesos sísmicos del Volcán Iraza, Rev. Geol. América Central, (1998). , 21, 47-59.

[33] Fernandez, M. Seismicity of the Pejibaye-Matina, Costa Rica, region: a strike-slip tectonic boundary?, Geofisica Internacional, 48(4), 351-364, (2009).

[34] Fisher, D, Gardner, T, Marshall, J, & Montero, W. Kinematics associated with late Cenozoic deformation in central Costa Rica: Western boundary of the Panama microplate. Geology, 22, 3: 263-266, (1994).

[35] Fisher, D. M, & Gardner, T. W. Tectonic escape of the Panama microplate: Kinematics along the western boundary, Costa Rica: Geological Society of America, Abstracts with Programs, (1991). , 23, A198.

[36] Fuh, S, Liu, C, Lundberg, N, & Reed, D. Strike-slip faults offshore southern Taiwan: implications for the oblique arc-continent collision processes, Tectonophysics, (1997). , 274, 25-39.

[37] Gardner, T. W, Fisher, D. M, & Marshall, J. S. Western boundary of the Panama microplate, Costa Rica: Geomorphological and structural constraints: International Association of Geomorphologists, 3rd International Geomorphology Conference,

August 23-28, 1993, McMaster University, Hamilton, Ontario, Canada, Programme with Abstracts, (1993). , 143.

[38] Goes, S. D. B, Velasco, A. A, Schwartz, S, & Lay, T. The April 22, 1991, Valle de la Estrella, Costa Rica (Mw=7.7) earthquake and its tectonic implications: a broadband seismic study, J. Geophys. Res., (1993). , 98, 8127-8142.

[39] Güendel, F, & Pacheco, J. The 1990-1991seismic sequence across central Costa Rica: evidence for the existence of a micro-plate boundary connecting the Panama deformed belt and the Middle America trench, Eos Trans. Am. Geophys. Un. 73, 399, (1992).

[40] Güendel, F, & Protti, M. Sismicidad y Sismotectónica de América Central, en: Buforn, E., Udías, A., Física de la Tierra, N° 10, Servicio de Publicaciones, Universidad Complutense de Madrid, (1998).

[41] Havskov, J, & Ottemøller, L. The SEISAN earthquake analysis software for Windows, Sun and Linux. Manual and software, Instutute of Solid Earth Physics, University of Bergen, Norway, (1999).

[42] Herrmann, R. B, Withers, M, & Benz, H. The April 18, 2008 Illinois earthquake: an ANSS monitoring success. Seism. Res. Lett., (2008). , 79, 830-843.

[43] Husen, S, Kissling, E, & Quintero, R. Tomographic evidence for a subducted seamount beneath the Gulf of Nicoya, Costa Rica: The cause of the 1990 Mw = 7.0 Gulf of Nicoya earthquake. Geophysical Research Letters, N 8, (2003). , 29

[44] Jacob, K, Pacheco, J, & Santana, G. Seismology and Tectonics, in Costa Rica Earthquake of April 22, 1991. Reconnaissance Report, Earthquake Spectra, Supplement B, (1991). , 7, 15-33.

[45] Kolarsky, R. A, Mann, P, & Montero, W. Island arc response to shallow subduction of the Cocos Ridge, Costa Rica, in Mann, P., ed., Geological Society of America Special Paper, Geologic and Tectonic Development of the Caribbean Plate Boundary in Southern Central America, (1995). (295), 235-262.

[46] Laurent, J. Evaluación económica de pérdidas y daños. 2009. En: Barquero (Ed.): El terremoto de Cinchona, 8 de enero de 2009. Inf. RSN, 101-127, (2009).

[47] Linkimer, L. Neotectónica del extremo oriental del Cinturón Deformado del Centro de Costa Rica, Tesis de Licenciatura, Universidad de Costa Rica, 103 , 2003.

[48] López, A. Neo and paleostress partitioning in the SW corner of the Caribbean plate and its fault reactivation potential. Tesis doctoral, Universidad de Tûbinger, Alemania, 293 , 1999.

[49] Lundgren, P, Protti, M, Donnellan, A, Heflin, M, Hernandez, E, & Jefferson, D. Seismic cycle and plate margin deformation in Costa Rica: GPS observations from 1994 to 1997, Journal of Geophysical Research, (1999). , 104(B12), 28915-28926.

[50] Lundgren, P, Wolf, S, Protti, M, & Hurst, K. GPS meaSurements of cristal deformation associated with the April 22, Valle de la Estrella, Costa Rica earthquake. Geophys. Res. Letters, (1993). , 20(5), 407-410.

[51] Mann, P, Schubert, C, & Burke, K. Review of the Caribbean neotectonic, in The Caribbean region, The Geology of North America, vol., H, edited by G. Dengo, and J. Case, Geological Society of America, Boulder, Colorado, (1990). , 375-391.

[52] Marshall, J, & Anderson, R. Quaternary uplift and seismic cycle deformation, Península de Nicoya, Costa Rica. GSA Bulletin, (1995). , 107(4), 463-473.

[53] Marshall, J. S. Evolution of the Orotina debris fan, Pacific coast, Costa Rica: Late Cenozoic tectonism along the western boundary of the Panama microplate: Geological Society of America, Abstracts with Programs, (1994). , 26(7), A207.

[54] Marshall, J. S, Fisher, D. M, & Gardner, T. W. Central Costa Rica deformed belt: Kinematics of diffuse faulting across the western Panama block, Tectonics, (2000). , 19, 468-492.

[55] Marshall, J. S, Fisher, D. M, & Gardner, T. W. Western margin of the Panama microplate, Costa Rica: Kinematics of faulting along a diffuse plate boundary: Geological Society of America, Abstracts with Programs, (1993). , 25(6), A284.

[56] Marshall, J. S, Gardner, T. W, & Fisher, D. M. Active tectonics across the western Caribbean-Panama boundary and the subducting rough-smooth boundary, Pacific coast, Costa Rica: Geological Society of America, Abstracts with Programs, (1995). , 27, A124.

[57] Marshall, J. S. LaFromboise, E.J., Utick, J.D., In the wake of flat subduction: Upperplate tectonics across a steep to flat slab transition, Pacific margin, Costa Rica, Central America: Backbone of the Americas, Patagonia to Alaska, 3-7 April 2006, Mendoza, Argentina, GSA Specialty Meetings Abstracts with Programs, Abs. 3-12, (2006). (2), 38.

[58] Matumoto, T, Othake, M, Lathan, G, & Umaña, J. Crustal structure of southern Central America, Bull. Seismol. Soc. Am., 67: 1:121-134, (1977).

[59] Ministerio de Industria, Energía y Minas (MIEM). Dirección de Geología, Minas y Petróleo, Mapa geológico de Costa Rica. Escala 1:200.000. San José, Costa Rica, (1982).

[60] Ministerio del Ambiente, Energía y Minas (MINAE). Dirección Superior de Geología, Minas e Hidrocarburos, Mapa geológico de Costa Rica. Escala 1:500.000. San José, Costa Rica, (1997).

[61] Montero, W, Neotectonics and related stress distribution in a subduction- collisional zone: Costa Rica, Profil: Stuttgart, (1994). , 125-141.

[62] Montero, W, Camacho, E, Espinosa, A, & Boschini, I. Sismicidad y marco neotectóni-
co de Costa Rica y Panamá. Rev. Geol. Amér. Central, v. espec., terremoto de Limón,
(1994). , 73-82.

[63] Montero, W, & Dewey, J. W. Shallow-focus seismicity, composite focal mechanism,
and tectonic of the Valle Central de Costa Rica. Seis. Soc. Amer. Bull, (1982). , 72

[64] Montero, W. El sistema de falla Atirro-Río Sucio y la cuenca de tracción de Turrialba-
Irazú: Indentación tectónica relacionada con la colisión del levantamiento del Coco,
Rev. Geol. Amer. Centr., (2003). , 28, 05-29.

[65] Montero, W. El terremoto del 4 de marzo de 1924 (Ms 7,0): ¿un gran temblor interpla-
ca relacionado al límite incipiente entre la placa Caribe y la microplaca Panama. Rev
Geológica de Amer. Central, (1999). , 22, 25-62.

[66] Montero, W, & Morales, L. D. Zonificación sísmica del Valle Central. Memorias del
4_ Seminario de Ingeniería Estructural, San José, CR, (1988).

[67] Montero, W, & Morales, L. Sismotectónica y niveles de actividad de microtemblores
en el suroeste del Valle Central, Costa Rica, Revista Geofísica, 21: 21-41, (1984).

[68] Montero, W. Neotectonica de la región central de Costa Rica: frontera oeste de la microplaca Panama. Rev. Geológica de Amer. Central, (2001). , 24, 29-56.

[69] Montero, W. Niveles de actividad de microtemblores en el sureste del Valle Central,
Costa Rica, Revista Geofísica 10-11: 105-115, (1979).

[70] Norabuena, E, Dixon, T, Schwart, S, Deshon, H, Newman, A, Protti, M, Gonzalez, V,
Dorman, L, Flueh, E, Lundgren, P, Pollitz, F, & Sampson, D. Geodetic, and seismic
constraints on some seismogenic zone processes in Costa Rica, J. Geophys. Res.
B11403, 1-25, (2004). , 109

[71] Pacheco, J, Quintero, R, Vega, F, Segura, J, Jiménez, W, & González, V. The Mw 6.4
Damas, Costa Rica, Earthquake of 20 November 2004: Aftershock and Slip Distribution, Bull. Seism. Soc. America N 4, doi:(2006). , 96

[72] Protti, M, Guendel, F, & Mcnally, K. The geometry of the Wadati-Benioff zone under
southern Central America and its tectonic significance: results from a high-resolution
local seismographic network, Phys. Earth Planet. Inter., (1994). , 84, 271-287.

[73] Protti, M, Schwarts, S. Mechanics of back arc deformation in Costa Rica: Evidence
from an aftershock study of the April 22, 1991, Valle de la Estrella, Costa Rica, earthquake (Mw = 7.7). Tectonics, N. 5: 1093-1107 , 13, 1994.

[74] Protti, M, Schwartz, S, & Zandt, G. Simultaneous inversion for earthquake location
and velocity structure beneath central Costa Rica, Seis. Soc. Amer. Bull., (1996). ,
86(1A), 19-31.

[75] Protti, M. The Most Recent Large Earthquakes in Costa Rica (1990 Mw 7.0 and 1991
Mw 7.6) and Three-dimensional Crustal and Upper Mantle P-wave Velocity Struc-

ture of Central Costa Rica, Ph.D. dissertation, University of California, Santa Cruz, 116, 1994.

[76] Quintero, R, & Guendell, F. Stress Field in Costa Rica, Central America, Journal of Seismology, (2000)., 4, 297-319.

[77] Ranero, C, & Von Huene, R. Subduction erosion along the Middle America convergent margin, Nature (2000)., 404, 748-752.

[78] Rojas, W. Catálogo de sismicidad histórica y reciente en América Central: Desarrollo y Análisis. Tésis de Licenciatura en Geología, Universidad de Costa Rica, 91, 1993.

[79] Rojas, W, Montero, W, Soto, G. J, Barquero, R, Boschini, I, Alvarado, G. E, & Vargas, A. Contexto geológico y tectónico local, sismicidad histórica y registro sísmico instrumental, In: Barquero, R. (Ed.): El terremoto de Cinchona, 8 de enero de 2009. Inf. Interno RSN: (2009)., 7-33.

[80] Seyfried, H, Astorga, A, Hubert, A, Calvo, C, Wolfgang, K, Hannlore, S, & Jutta, W. Anatomy of an evolving Island Arc: tectonic and eustatic control in the south Central American forearc area, in: McDonald, D.I.M (Ed.): Sea level Changes at active plate margins: Processes and Products. Spec. Publs. Int Assoc. Sediments, (1991)., 12, 217-240.

[81] Snoke, J. A, Munsey, J, Tiague, W, & Bollinger, A. C. G. A., a program for focal mechanism determinations by combined use of polarity and SV-P amplitude ratio data, earthquakes, 55(3): 15., (1984).

[82] Soulas, J. Tectonica activa, informe de mision de consultuoria P. H. Siquirres, Instituto Costarricense de Electricidad (ICE), Internal report, (1989).

[83] Stoiber, R, & Carr, M. Quaternary volcanic and tectonic segmentation of Central America: Bull. Volc. (1973)., 37(3), 304-323.

[84] Suárez, G, Pardo, M, Domínguez, J, Ponce, L, Montero, W, Boschini, I, & Rojas, W. The Limón, Costa Rica, earthquake of April 22, 1991: Back arc thrusting and collisional tectonics in a subduction environment. Tectonics, (1995)., 14(2), 518-530.

[85] Tournon, J, & Alvarado, G. Carte géologique du Costa Rica: notice explicative; Mapa geológico de Costa Rica: folleto explicativo, échelle-escala 1 500 000.-Ed. Tecnológica de Costa Rica, 80 pp. + Mapa geológico de Costa Rica, (1997).

[86] Trenkamp, R, Kellog, J, Freymueller, J, & Mora, H. Wide plate margin deformation, southern Central America and Northwestern South America, CASA GPS observations, Journal of South American Earth Sciences, (2002)., 15, 157-171.

[87] Van Andel, T. H, Heath, G. R, Malfait, B. T, Heinrichs, D. F, & Ewing, J. I. Tectonics of the Panama Basin, eastern equatorial Pacific. Geological Society of America Bulletin, (1971)., 82, 1489-1508.

[88] Von Huene, R, Ranero, C, & Watts, P. Tsunamigenic slope failure along the Middle America Trench in two tectonic settings. Marine Geology, (2004). , 203, 303-317.

[89] White, R, & Harlow, D. Destructive Upper-Crustal Earthquakes of Central America Since 1900. Bull. Seims. Soc. Am., (1993). , 83

[90] White, R. Tectonic inplications of upper-crustal seismicity in Central America, In: Slemmons, D., Engdahl, E., Zoback, M., Blackwell, eds, Neotectonics of North America, Boulder Colorado, Geological Society of America, Decade Map (1991). , 1

[91] Woodward-Clyde: A preliminary evaluation of earthquake and volcanic hazards significant to the major populations centers of the Valle CentralCosta Rica. Final Report prepared for Ret Corporation, San José, Costa Rica, (1993).

[92] Yao, Z, Quintero, R, & Roberts, R. Tomographic Imaging of P- and S- wave velocity structure Veneta Costa Rica. Journal of Seismology (1999). , 3, 177-190.

[93] Zhu, L, & Helmberger, D. Advancement in source estimation technique using broadband regional seismograms. Bull. Seism. Soc. Am., (1996). , 86, 1634-1641.

Chapter 5

Modeling Dynamic-Weakening and Dynamic-Strengthening of Granite in High-Velocity Slip Experiments

Zonghu Liao and Ze'ev Reches

Additional information is available at the end of the chapter

1. Introduction

Earthquakes are associated with slip along fault-zones in the crust, and the intensity of dynamic-weakening is one of the central questions of earthquake physics (Dieterich, 1979; Reches and Lockner, 2010). Since it is impossible to determine fault friction with seismological methods (Kanamori and Brodsky, 2004), the study of fault friction and earthquake weakening has been usually addressed with laboratory experiments (Dieterich, 1979) and theoretical models (Ohnaka and Yamashita, 1989).

The experimental analyses of dynamic-weakening were conducted in several experimental configurations: bi-axial direct shear (Dieterich, 1979; Samuelson et al, 2009), tri-axial confined shear (Lockner and Beeler, 2002), and rotary shear apparatus (Tsutsumi and Shimamoto, 1997; Goldsby and Tullis, 2002; Di Toro et al., 2004; Reches and Lockner, 2010). The direct shear apparatus allows high normal stress and controlled pore water pressure (up to ~200 MPa) with limited slip velocity (up to 0.01 m/s) and limited slip distance (~10 mm) (Shimamoto and Logan, 1984). These slip velocities and displacements are significantly smaller than those of typical earthquakes (0.1-10 m/s and up to 5 m, respectively). In order to study high velocity and long slip distance, experiments have been conducted in rotary shear machines.

While many studies indicated a systematic weakening with increasing slip-velocity (Dieterich, 1979; Di Toro et al., 2011), recent experimental observations revealed an opposite trend of dynamic-strengthening particularly under high velocity (Reches and Lockner, 2010; Kuwano and Hatano, 2011). This strengthening was attributed to dehydration of the fault gouge due to frictional heating at elevated velocities (Reches and Lockner, 2010; Sammis et al., 2011). If similar dynamic-strengthening occurs during earthquakes, it should be incorporated in the analyses of earthquake slip.

We analyze here of the relations between friction coefficient and slip velocity for steady-state velocity the range of 0.001-1 m/s. These relations are derived for the Sierra White granite (SWG) experiments under normal stress up to 7 MPa. The results are presented by a model that is termed WEST (WEkening-STrengtheing). We first present the main experiment observations, for five rock types, and then we derive the numeric model for Sierra White granite (SWG). Finally, we apply the numerical model to tens of experiments with complex velocity history conducted on the same rock.

2. Experimental observations

The experiments were conducted with a rotary shear apparatus and cylindrical rock samples (Reches and Lockner, 2010; Chang et al., 2012). The experimental setup and sample configuration are described in Appendix A. The experimentally monitored parameters include slip distance, slip velocity, fault-normal displacement (dilation), shear stress and normal stress. Sample temperature was measured by thermocouples that are embdded ~3mm away from the slip surfaces. The normal load was maintained constant during a given experiment, and the experiments were performed at room conditions. Fault strength is represented by the friction coefficient, $\mu = \tau/\sigma_N$ (where τ is shear stress and σ_N the normal stress). The slip velocity was either maintained constant, or increased or decreased in steps. We present results of tests with samples of Sierra White granite (after Reches and Lockner, 2010), and new results for samples of Blue quartzite, St. Cloud diorite, Fredricksburg syenite, and Karoo gabbro. The tests with SWG samples were run under the widest range of conditions and the numeric model was derived only for this rock.

Reches and Lockner (2010) determined the friction-velocity relations of SWG for a velocity range of.0003-1 m/s and normal stress up to 7 MPa. Their results revealed three general regimes of friction-velocity relations (Fig. 1a):

i.   Dynamic-weakening (drop of 20-60% of static strength) as slip velocity increased from ~0.0003 m/s to a critical velocity of $V_c \sim 0.03$ m/s, during which the friction coefficient was 0.3-0.45.

ii.   Transition to dynamic-strengthening regime in the velocity range of V = 0.06-0.2 m/s, during which the fault strength almost regained its static strength; and

iii.   Quasi-constant strength for V > 0.2 m/s, with possible further drops as velocity approaches ~1 m/s. Only few experiments were conducted in this range due to sample failure by thermal fracturing.

Similar pattern of weakening-strengthening of Westerly granite samples was recently observed by Kuwano and Hatano (2011) (Fig. 1b). They showed that the friction coefficient dropped in the velocity range of 0.001-0.06 m/s, and rose in the velocity range of 0.06-0.2 m/s

(Fig. 1b). Earlier work of Tsutsumi and Shimamoto (1997) also indicated temporal periods of strengthening, in which the friction sharply increased, and visible melting was observed at the strength peak when the slip rate was increased from 0.55 to 0.73 m/s.

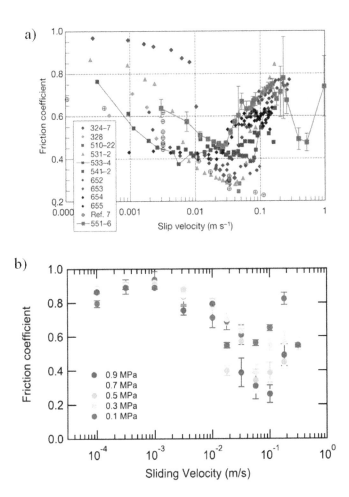

Figure 1. Experimental friction-velocity relations in (a) Sierra White granite (after Reches and Lockner, 2010) and (b) Westerly ganite (after Kuwano and Hatano, 2011

To further explore the occurrence of dynamic-strengthening, we tested four other rocks under conditions similar to Reches and Lockner (2010) tests. The new experiments were conducted on samples made of Blue quartzite (Fig. 2a), St. Cloud diorite (Fig. 2b), Fredricksburg syenite (Fig. 2c), and Karoo gabbro (Fig. 2d). The diorite (Fig. 2b) and syenite (Fig. 2c) samples displayed a distinct transition into dynamic-strengthening regime (red arrows). The critical transition velocity, V_C, depends on the sample lithology; it is ~0.02 m/s and ~0.01 m/s for the diorite and syenite, respectively. The Blue quartzite tests displayed only negligible weakening and strengthening (Fig. 2a).

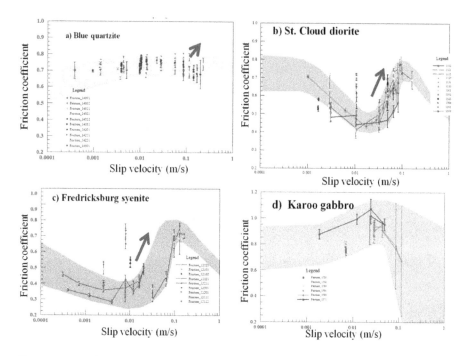

Figure 2. Friction-velocity relations in shear experiements of Blue quartzite (a), St. Cloud diorite (b), Fredricksburg syenite (c), and Karoo gabbro (d). Connected lines indicated data from step-velocities experiments. Unconnected dots indicate single velocity experiments.Note velocity-weakening and velocity-strengthening stages (marked by red arrows) in (a)-(c) and no velocity-strengthening in (d).

3. Numerical modeling

3.1. Approach

The most striking feature of the relations between the friction coefficient and slip-velocity for the above experiments is the systematic transition from weakening under low velocity to strengthening at higher velocities (Figs. 1, 2). We refer to this feature of **WE**akening-**ST**rengthening as the **WEST** mode, and developed a numerical model to describe its character. The model is based on two central assumptions. First, the friction coefficients, in both weakening and strengthening regimes, can be presented as dependent of the slip distance, and slip velocity. This assumption was employed in many previous models (Dieterich, 1979; Beeler et al, 1994; Tsutsumi and Shimamoto, 1997; Reches and Lockner, 2010; Di Toro et al., 2011). Second, the weakening-strengthening mode reflects a transition between different frictional mechanisms. Reches and Lockner (2010) and Sammis et al. (2011) suggested that the weakening is controlled by gouge powder lubrication due to coating of the powder grains by a thin layer of water that is 2-3 monolayers thick. The dehydration of this layer at elevated temperature under high slip velocity leads to strengthening. It is thus assumed that the weakening and strengthening regimes have different parametric relation between friction and velocity.

3.2. Model formulation

Following the above assumptions, we model the relations of the steady-state friction coefficient, μ, and the slip distance, D, and slip-velocity, V (Fig. 3). The steady-state is defined for a given slip velocity during which the weakening (or strengthening) intensifies with increasing slip distance. The model addresses the three major features of the experimental observation: (I) the slip-weakening relation, $\mu(V, D)$, represents the drop from static friction coefficient, μ_S, to the kinematic friction coefficient, μ_K during slip to the critical distance, D_C, at a constant velocity; (II) the dynamic-friction coefficient, $\mu_K(V)$, at $V < V_C$, under steady-state; and (III) the dynamic-strengthening at high velocity regime of $V > V_C$. We regard the friction coefficient decrease from (1) to (2) (Fig. 3) as the dynamic-weakening, and the friction coefficient increase from (2) to (3) as the dynamic-strengthening.

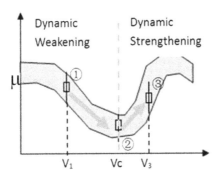

Figure 3. A schematic presentation of WEST model, with velocity controlled kinematic friction coeefcient.

The dependence of the experimentally monitored friction coefficient, μ, on the two experimentally controlling factors of D and V has the general form of

$$\mu = \mu\left(D, V\right) \tag{1}$$

The friction coefficient, μ, has two end members, the static friction coefficient, μ_S, which is the friction coefficient during slip initiation, and the kinematic friction coefficient, μ_K, the steady-state coefficient after large slip distance under a constant velocity. The transition between the weakening regime and the strengthening regime occurs at the critical slip-velocity, V_C, that is determined from the general plot of friction coefficient as function of slip-velocity (Fig. 1, 2, 3).

We assume that during the weakening stage and under constant velocity, the drop of μ from μ_S to μ_K is controlled only by the slip distance D,

$$\mu = \mu(D), when \mu_S > \mu > \mu_K, \tag{2}$$

where,

$$\mu = \mu_S \ for \ V \ = \ 0 \ and \ D \ = \ 0 \ \left(Slip \ initiation\right), \qquad a$$
$$and \tag{3}$$
$$\mu = \mu_K \ for \ steady-state \ velocity, \ V, \ and \ D \ > \ D_C, \qquad b$$

where D_C is the critical slip distance.

It is also assumed that μ_K is a function only of slip velocity V,

$$\mu_K = \mu_K(V), \tag{4}$$

Similarly, we assume that during the strengthening stage, μ is a function of both slip-velocity and slip distance,

$$\mu = \mu(D, V), for \ V \ > \ V_C, \tag{5}$$

This approach for the empirical relations is in the spirit of the weakening relations developed by Dieterich (1979) who introduced the rate- and state- friction law.

3.3. Model parameterization

General

Solutions for the model functional relations were searched in the following steps.

1. Prepare a table of the experimental velocity-friction relation;
2. Select experiments with representative weakening stage;
3. Search for a suitable functional fit between slip velocities, friction coefficient, and slip distance. For this search, we chose the simplest functional relation that provided good fit. The search revealed the relations between the μ_K and velocity.
4. Integrate the above relations for the effect of distance and velocity.

5.   Apply the above functional relations to set up the specific WEST model (Equations 2, 4, 5).

Dynamic-weakening regime

The experimental data set has large data scatter at low velocities (Fig.1; Reches and Lockner, 2010). For the range of V < 0.03 m/s (Fig. 4), we found the following relations of μ_K and slip velocity,

$$\mu_K(V)=0.742 - \frac{0.375V}{0.00183 + V}, \text{ for } V \le 0.03 \text{ m}/\text{s} \tag{6}$$

The RMS (root mean square) of this solution is 0.83 while the correlation coefficient is 0.73. This simple solution provides reasonable fit to the scattered friction data (Fig. 4).

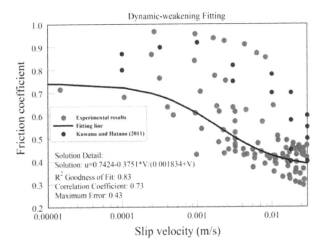

Figure 4. The selected solution (red curve) for dynamic-weakening of kinetic friction coefficient of SWG for V = 0.0003-0.03m/s by Eureqa. Data for V < 0.003 m/s are from Reches and Lockner (2010) and Kuwano and Hatano (2011).

Dynamic-strengthening regime

In the strengthening regime of 0.03 m/s < V < 1.0 m/s (Fig. 5), the selected solution for velocity-controlled friction coefficient is

$$\mu_K(V)=0.824 \exp\left(-\frac{0.0275}{V}\right) \text{for } V>0.03m/s \tag{7}$$

The RMS of this solution is 0.91 and the correlation coefficient is 0.74. The trend of this exponential relation provide good fit for SWG (Reches and Lockner, 2010) in the strengthening trend (Fig. 5), and this trend generally fit the experimental results of Kuwano and Hatano (2011). Note: Friction during 0.305-0.4 m has been locally adjusted to comply with Fig. 4.

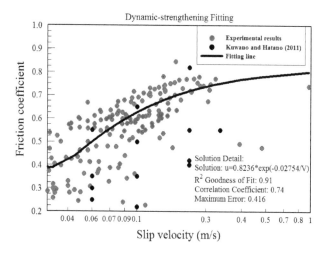

Figure 5. a) The selected solution (red curve) for dynamic-strengthening of SWG for V = 0.03-0.3m/s by Eureqa. Data for V > 0.003 m/s are from Reches and Lockner (2010) and Kuwano and Hatano (2011)

Slip-weakening relations

For the ten experimental results (Fig. 6), we selected the slip-weakening function as,

$$\mu = \mu(D) = 0.2663\ (\pm 0.1)\exp(-D) + 0.4061(\pm 0.15) \tag{8}$$

For D=0, which is slip initiation, the calculated $\mu = \mu_S = 0.6724$. For D > Dc (e.g., 3.0 m), the calculated $\mu = \mu_K = 0.41$.

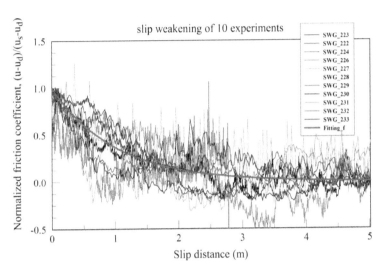

Figure 6. a) The best fit solution (red curve) for slip distance weakening in ten experiments. SWG 222, 223, 224 were run at velocity 0.024 m/s; SWG 226-232 were run at 0.072 m/s; σ_N = 1.1 MPa. The solution is μ= 0.2663(±0.1) exp (-D) +0.4061(±0.15). The constants can be adjusted for specific case

Model synthesis

By substitute the 0.4061 (Eq. 8) by Eq. (7), Eq. (8) is generalized into,

$$\mu(D, V) = 0.2663 \bullet \exp(-D) + \mu_K(V) \tag{9}$$

which combines Eq. (2, 4, 5). This simple exponential function characterized the weakening process with relating small deviation of the data. A summary of WEST equations is listed in Table 1.

Terms	Expression	Eq. Number
General function	$\mu = \mu(D, V)$	(1)
Slip-weakening	$\mu(D, V) = 0.2663 \exp(-D) + \mu_K(V)$	(9)
Dynamic-weakening	$\mu_K(V) = 0.742 - \frac{0.375V}{0.00183 + V}$, for $V \leq 0.03$ m/s	(6)
Dynamic-strengthening	$\mu_K(V) = 0.824 \exp\left(-\frac{0.0275}{V}\right)$, for $V > 0.03 m/s$	(7)

Table 1. A summary of WEST equations.

4. Model application and analysis

4.1. General

We now apply the WEST model to simulate the friction-velocity-distance evolution in a set of experiments with Sierra White granite. In these simulations, we use the experimental slip distance, D, and slip velocity, V, which were controlled by the operator, as input parameters in Equations (6, 7, 9) and Table 1. This substitution predicts the evolution of the friction coefficient during the experiments. The predicted friction coefficient is then plotted with the experimentally observed equivalent. The simulations were done on four types of velocity histories; none of the simulated experiment was used for the derivation of the model.

4.2. Rising velocity steps

We start with three runs under same σ_N = 5.0MPa, and each with three upward stepping velocities, 0.0025, 0.025 and 0.047 m/s. During these runs the cumulative slip was about 9 m (Fig. 7). These velocities are in the weakening regime, and they display intense friction drops that reflect both the rising velocity and increase of slip distance. The friction coefficient was 0.7-0.9 at the first step of V = 0.0025 m/s, and it decreased to $\mu \cong 0.5$ in the second velocity step of V = 0.025 m/s. A minimum of friction coefficient of μ = 0.4 developed under V = 0.05 m/s. This experimental evolution fits well the predictions by WEST model.

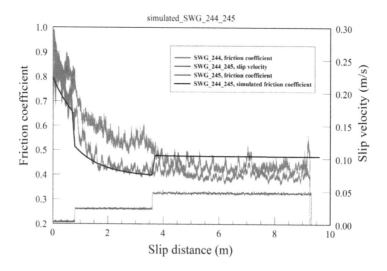

Figure 7. WEST model simulation and experimental results for SWG experiments 244, 245, SWG, σ_N = 5.0 MPa

4.3. Rise and drop

We now examine two groups of experiments, each with five runs of the same loading conditions (σ_N = 5.0MPa). Each group has three velocity stages: low, high, and back to low. In the first group, the three steps were 0.00075, 0.0075, and 0.00075 m/s. The initial friction coefficient ranged 0.4-0.7 for the different runs in the group, and during the first step, the friction coefficient was gradually decreased by ~0.1. During the second stage of V = 0.0075 m/s, which lasted 10 s, the friction coefficient reduced drastically, with larger reduction of the runs that started at higher friction coefficient. During the final, low velocity, the sample was slightly strengthened.

In Fig. 8, this group of experiments was performed on the same sample under σ_N = 5.0MPa. The different initial friction coefficients (e.g., runs 1, 2 have higher friction coefficient than run 3, 4, 5) were attributed to the holding-times between experiments. The weakening was well noticed when slip velocity was raised and then the friction reached a relative steady value. The fit line shows the weakening immediately after the velocity has been raised. When the velocity drops, the friction coefficient is raised higher in the fit line than the data.

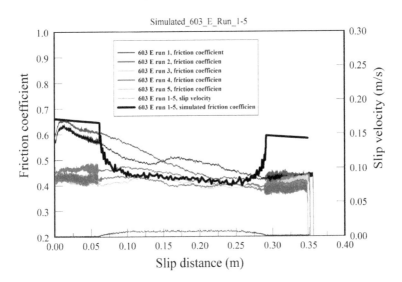

Figure 8. WEST model simulation and experimental results for SWG_603_Run_(1-5), SWG, σN = 5.0 MPa

In the second group (Fig. 9), the scenario was repeated but at velocities which were higher by an order of magnitude: 0.0075 m/s (first and last stages) and 0.075 m/s (second stage). This time, the friction coefficient in the second stage was increased from μ=~0.3 to the highest μ = 0.7. Then, μ gradually reduced to 0.3-0.4 during the final, low velocity stage. In Fig. 9, the friction coefficient of the fit results rises and drops earlier than the experimental data (running from 0.25 m to 0.75 m). The rest of the running, at the beginning low velocity and the final velocity, the friction coefficient between the model and experiments fit well each other.

Comparing the two groups, the different friction responses provide clear evidence for dynamic-strengthening at slip velocity ~ 0.07 m/s.

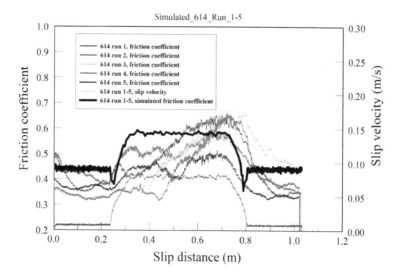

Figure 9. WEST model simulation and experimental results for SWG_614_Run_(1-5), SWG, σ_N = 5.0 MPa

The WEST simulations were also performed to investigate if the model can capture the dynamic-strengthening at the high velocity regime. Fig. 10 is an experiment with five alternating steps of velocity that again indicates the dynamic-strengthening at velocity 0.075 m/s (above the critical value V_C). The friction coefficient dropped from 0.7 to 0.5 at velocity of 0.0025 m/s and 0.025 m/s, then dynamically-strengthened to $\mu = 0.7$ at $V = 0.75$ m/s, and finally slightly dropped to 0.55 with a slip distance of ~4.2 m (from 1.7 m to ~5.9 m) as the velocity decreased. Similarly to the case of rise-drop experiment (Fig. 8), the observed friction coefficient changed only slightly in the end at dropping velocity (from 5.9 m to 7.6 m). The simulation illustrates the dynamic-strengthening during the three-step sliding.

4.4. Drop and rise during long distance slip experiments

In Fig. 11, the simulation presents the experiment behavior that involved three steps: an initial weakening at the higher slip velocity of 0.045 m/s, regaining strengthen at a lower speed of 0.0018 m/s, and a final weakening at high-speed step of 0.045 m/s after healing for about 2,000 seconds. The healing refers to the regaining of strength of the fault during hold time. During the simulation of the second step, the friction coefficient was manually shifted to static value due to healing (indicated by blue arrow, Fig. 11). The increase of friction here is related to healing during hold time, which differs from dynamic-strengthening. The simulation results show smooth lines instead of noisy curves of the experiment.

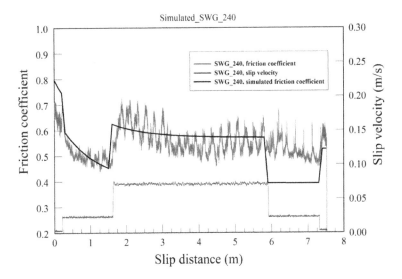

Figure 10. WEST model simulation and experimental results for SWG 240, SWG, σ_N = 1.1 MPa

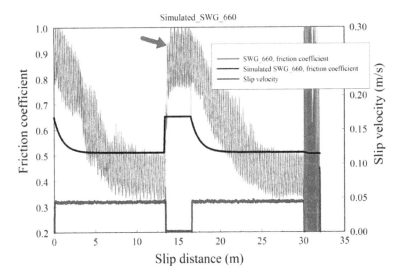

Figure 11. WEST model simulation and experimental results for SWG 660, SWG, σ_N = 0.5MPa. The rising of friction coefficient indicated by blue arrow is NOT dynamic-strengthening

4.5. Wide velocity range

Figs. 12 and 13 display the friction evolution in two experiments with wide velocity range. Fig. 12 displays a slide-hold-slide run under σ_N = 1.1 MPa and hold times of 10 s between the multiple velocity steps. Under this short hold time, the friction coefficient curve displays quasi-continuous trend. There is a marked weakening as the velocity was increased to ~ 0.04 m/s, followed by a gentle strengthening at higher velocities. The modeling results were lower at beginning than expected and strengthening occurred earlier. Fig. 13 displays two major features: an initial gradual weakening in the slip velocity range of ~0.0003 m/s to a critical velocity of ~0.03 m/s, and a fast strengthening at velocities from ~0.03 m/s to 0.2 m/s. In the final stage, the friction reaches ~0.8. The modeling simulates the weakening-strengthening pattern, but it fails to follow the experimental results in the region faster than the critical velocity. In the experiment, the friction coefficient remained relatively low from ~3 m to ~11 m, where the simulated results predicted earlier strengthening. Also, an abrupt friction rise was observed in the experiment (at ~ 11 m) whereas the model indicated smooth and continuous strengthening.

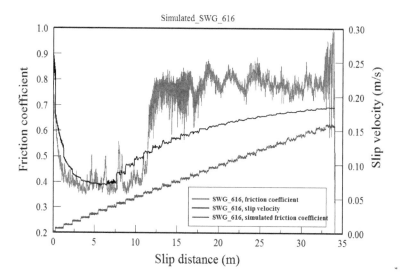

Figure 12. WEST model simulation and experimental results for SWG 531, SWG, σ_N = 1.1MPa. Friction coefficient is shown for a full-velocity by interval holding in sliding distance together with a best fit modeling result

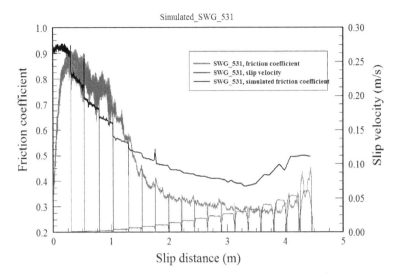

Figure 13. WEST model simulation and experimental results for SWG 616, SWG, σ_N = 5.0 MPa. Friction coefficient is shown for a full-velocity continuously in sliding distance together with a best fit modeling result

5. Discussion

5.1. Field applications

The WEST model was developed for the experimental results of Sierra White granite for which we have multiple runs that well bound the solutions and simulations. The WEST relations (Eqn. 6, 7, 9) represent the empirical frictional history of SWG in the slip-velocity range of 10^{-4}-1m/s. The application of the model to the experimental results shows that friction history can be captured by a numerical combination of the slip velocity and slip distance for both the weakening and strengthening stages. Specifically, in the dynamic-strengthening regime has the following properties: 1) It is a high-velocity regime, $V > V_C$, e.g. $V \sim 0.03$-0.6 m/s for Sierra White granite; 2) The WEST model uses a kinematic friction coefficient for the strengthening stage, which is derived independently of the weakening stage; 3) The friction coefficient history is well simulated by combining the kinematic friction coefficient for strengthening and the slip distance.

The WEST model predicts the rate-dependence of friction over the full range of observed natural seismic slip-rates, and in this sense it is applicable to earthquake simulations, and modeling of earthquake ruptures. Accumulating evidence supports the presence of dynamic-strengthening. Kaneko et al (2008) stated that "the velocity-strengthening region suppresses supershear propagation at the free surface occurring in the absence of such region, which could explain the lack of universally observed supershear rupture near the free surface". Hence, it is important to understand how the dynamic friction at seismic rates affects earthquake rupture dynamics by adopting new friction model in the numerical simulations. Although this is beyond the scope of the present study, the abundance of efforts on experimental studies on dynamic friction provides a realistic velocity-friction relation to explore the ground motions during earthquake rupture.

6. Summary

The analysis examines the frictional strength of igneous rocks including granite, diorite, syenite, gabbro, and quartzite under slip-velocity approaching seismic-rates. The experimental observations confirm that increasing slip velocity leads to dynamic-weakening followed by **dynamic-strengthening** only in igneous rock samples that contain quartz (Figs. 1, 2). The weakening-strengthening transition occurs at a critical velocity, V_C, which depends on the fault lithology (Figs. 1, 2).

The present study provides a numerical framework for the experimental observations of dynamic-strengthening at high velocity along rock faults. The model is empirical in nature, and it succeeds to simulate the friction coefficient history in a wide range of experimental loading. We envision that it is a promising tool to analyze rock friction during earthquakes. In future research, several aspects of the sliding mechanism should be linked and interpreted in terms of the WEST model: 1) Powder lubrication by Reches and Lockner (2010) showed that the newly formed gouge organizes itself into a thin deforming layer that changes the fault strength; 2) Fault-strengthening during hold time; and 3) Predicting the effect of lithological compositions on the frictional resistance (Di Toro et al., 2004).

Appendix A: Experiment setup

The analyzed experiments were conducted on the ROGA in University of Oklahoma. The following description of the apparatus is taken from Reches and Lockner (2010) and related lab proposals. The ROGA system (Rotary Gouge Apparatus, Fig. 14) satisfies the following conditions: (1) normal stress of tens to hundreds of MPa; (2) slip velocity of ~1 m/s; (3) rise-time of less than 1 s; and (4) unlimited slip distances.

In the experiments, the fault is composed of solid blocks of Sierra White granite. Each sample includes two cylindrical blocks, diameter = 101.6 mm, height = 50.8 mm. For ~uniform velocity, the upper block has a raised ring with ID = 63.2 mm and OD = 82.3 mm; the blocks are pressed across this raised ring. Thermocouples are cemented into holes drilled 3 mm and 6 mm away from the sliding surfaces (Fig. 14). Each pair of blocks wore to form gouge in between at different slip velocities of 0.001-1 m/s. A large set of experiments will be executed and a large quantity of data will be collected by LabView at frequency of ~100-1000 Hz.

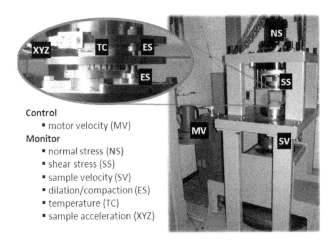

Figure 14. The Rotary Shear Apparatus with builder Joel Young. The sample block assembled in the loading frame (after Reches and Lockner (2010)

Acknowledgements

Discussions and suggestions by David Lockner and Yuval Boneh contributed to the analysis. Usage of the 'Eureqa Formulize' program helped in the model computation. Partial funding support was provided by grants 1045414 of NSF Geosciences, Geophysics, G11AP20008 of DOI-USG-NEHRP2011, and SCEC 2012.

Author details

Zonghu Liao and Ze'ev Reches

School of Geology and Geophysics, University of Oklahoma, Norman, USA

References

[1] Beeler, N. M., T. E. Tullis, and J. D. Weeks (1994), The roles of time and displacement in the evolution effect in rock friction, *Geophys. Res. Lett.*, 21(18), 1987-1990.

[2] Boneh, Y. (2012), Wear and gouge along faults: experimental and mechanical analysis, M.S. thesis, Dep. of Geol. and Geophys., Univ. of Oklahoma, Norman, OK, USA.

[3]   Chang, J. C., Lockner, D. A., and Z. Reches (2012), Rapid acceleration leads to rapid weakening in earthquake-like laboratory experiments, *Science*, 338(6103), 101-105.

[4]   Di Toro, G., D. L. Goldsby DL, and T. E. Tullis (2004), Friction falls towards zero in quartz rock as slip velocity approaches seismic rates, *Nature*, 427, 436-439.

[5]   Di Toro, G., R. Han, T. Hirose, N. De Paola, S. Nielsen, K. Mizoguchi, F. Ferri, M. Cocco, and T. Shimamoto (2011), Fault lubrication during earthquakes, *Nature*, 471, 494-498.

[6]   Dieterich, J. H. (1979), Modeling of rock friction: 1. Experimental results and constitutive equations, *J. Geophys. Res.*, 84(5), 2161-2168.

[7]   Goldsby, D. L., and T. E. Tullis (2003), Flash heating/melting phenomena for crustal rocks at (nearly) seismic slip rates, SCEC Annual Meeting Proceedings and Abstracts, Palm Springs, California.

[8]   Kanamori, H., and E. E. Brodsky (2004), The physics of earthquakes, Reports on *Progress in Physics*, 67, 1429 – 1496, doi: 10./1088/0034-4885/67/8/R03.

[9]   Kaneko, Y., N. Lapusta, and J.-P. Ampuero (2008), Spectral element modeling of spontaneous earthquake rupture on rate and state faults: Effect of velocity-strengthening friction at shallow depths, *Journal of Geophysical Research*, 113, B09317, doi: 10.1029/2007JB005553.

[10] Kuwano, O., and T. Hatano (2011), Flash weakening is limited by grannular dynamics, *Geophys. Res. Lett.*, 38, L17305, doi:10.1029/2011GL048530.

[11] Liao, Z. (2011), Dynamic strengthening at high-velocity shear experiments, MS thesis, University of Oklahoma, Norman, 54 pp.

[12] Liao, Z., and Z. Reches (2012), Experiment-based model for granite dynamic strength in slip-velocity range of 0.001-1.0 m/s, AGU Annual Meeting, San Francisco, 2012, T13E-2656, ID: 1488537.

[13] Lockner, D. A., and N. M. Beeler (2003), Stress-induced anisotropic poroelasticity response in sandstone, Electronic Proc. 16th ASCE Engin. Mech. Conf., Univ. of Washington, Seattle, WA.

[14] Marone, C. (1998), Laboratory-derived friction laws and their application to seismic faulting, *Annu. Rev. Earth Planet Sci.*, 26, 643-696.

[15] Ohnaka, M., and T. Yamashita (1989) A cohesive zone model for dynamic shear faulting based on experimentally inferred constitutive relation and strong motion source parameters, *Journal of Geophysical Research*, 94 (B4), 4089-4104.

[16] Reches, Z., and D. A. Lockner (2010), Fault weakening and earthquake instability by powder lubrication, *Nature*, 467(7314), 452-455, doi:10.1038/nature09348.

[17] Sammis, C. G., D. A. Lockner, and Z. Reches (2011), The role of adsorbed water on the friction of a layer of submicron particles, *Pure and Applied Geophysics*, doi: 10.1007/s00024-01-0324-0.

[18] Samuelson, J., D. Elsworth, and C. Marone (2009), Shear-induced dilatancy of fluid-saturated faults: experiment and theory, *J. Geophys. Res.*, 114, B12404, doi: 10.1029/2008JB006273.

[19] Schmidt, M., and H. Lipson (2009), Distilling free-form natural laws from experimental data, *Science*, 324, 81-85.

[20] Shimamoto, T., and J. M. Logan (1984), Laboratory friction experiments and natural earthquakes: An argument for long teren tests, *Technophysics*, 109, 165-175.

[21] Tsutsumi, A., and T. Shimamoto (1997), High-velocity frictional properties of gabbro, *Geophys. Res. Lett.*, 24(6), 699-702.

Chapter 6

Parameters Identification of Stochastic Nonstationary Process Used in Earthquake Modelling

Giuseppe Carlo Marano, Mariantonietta Morga and
Sara Sgobba

Additional information is available at the end of the chapter

1. Introduction

The design of structures resistant to seismic events is an important field in the structural engineering, because it reduces both the loss of lives and the economic damages that earthquakes can produce. The accuracy and the robustness of the design of structures resistant to seismic events are still not completely guaranteed. In order to define rules in the design codes to design earthquake-resistant structures, several scholars have investigated the probability of a seismic event to occur in a specific location and its characteristics, like the intensity and the return time (e.g. frequency). Indeed, the return time and the characteristics of the earthquakes occurring in a given area determine the dynamic loads exciting a structure built in that area for its whole lifetime. The structural response to ground motion is function of the seismological parameters of the area where the earthquake occurs and the structure is built, in addition to the kind of structure. The earthquake characteristics related to the seismological parameters that strongly influence the structural response are the earthquake intensity, the rupture type and the epicentral distance. This leads to define the seismic dynamic loads exciting a structure as function of these seismological parameters. Unfortunately, the seismological parameters are not very useful in structural design. Instead, peak amplitude, frequency content, energy content and duration of the event are the characteristics of the earthquakes useful to structural design.

To design strategic or complex structures and infrastructures resistant to earthquakes, the analysis of the dynamic time-history response of the structure to earthquake records is preferred to the response spectrum analysis. Indeed, the dynamic time-history response provides temporal information of the structural response that is essential in non-linear analysis of some kind of structures to estimate their level of damage. Some design codes indicate the use of real records of earthquake ground motions as input of the dynamic time-history analysis of the structural response. Unfortunately, the selection of natural earthquake accelerograms that adhere to some criteria, such as the response spectrum for some design scenarios, is difficult. Indeed, the number of occurred earthquakes recorded in a specific area or with some characteristics is often not sufficient, because a wide set of accelograms is required for the design of earthquakes-resistant structures. To overcome this difficulty some design codes allow the use of modified natural records (with changes either in the time domain or in frequency domain) or synthetic accelerograms in the dynamic time history-analysis of the structural response. Unlikely, other design codes preclude the use of artificial accelerograms for the dynamic time-history analysis of the structural response because of the difficulty to generate accelograms that adhere to criteria for some design scenarios [1]. The approach based on the natural accelerograms is prevailing, since a real recorded accelerogram properly processed is undeniably a realistic representation of the ground shaking that is occurred in a particular seismological scenario. On the other hand, the recorded accelerogram represents a past seismic event occurred in a specific area and not a future event that will occur in that area and cannot be predicted because of stochastic nature of the seismic ground motions. This is a further reason to generate artificial accelerograms for the structural design on the basis of a stochastic model.

Several scholars have proposed different methods to generate the synthetic accelograms, but nowadays no model is indicated in the design codes to generate the artificial records of seismic ground motions. Moreover, the design codes that allow the use of artificial accelograms prescribe that the mean response spectrum of the synthetic earthquake records has to match a given response spectrum within a given tolerance.

The complex nature of the release of seismic wave, their propagation in soil and the unpredictability of the earthquake occurrence make the stochastic-based approach the most suitable to model the earthquake ground motion. In that sense the earthquake occurring in a specific area is modelled as a stochastic process, so each recorded seismic ground motion is defined as a sample function of that stochastic process. The artificial accelerograms are also sample functions of the stochastic process modelling the earthquake occurring in that area: they represent the possible future seismic events. For this reason in the design phase of a complex structure the structural response to these artificial accelerograms is calculated.

The stochastic process modelling the seismic events occurring in an area is defined through the characteristics of the strong ground motions recorded in that area. Several scholars have presented methods to define the stochastic model to describe the seismic ground motion and simulate artificial earthquake records. Firstly, stationary filtered white noise model have been proposed to describe and simulate earthquakes [2, 3]. The most known of these models is the Kanai–Tajimi model [4, 5]. Some scholars have modified this model [6] or have proposed stationary multi-filtered white noise models, as the Clough-Penzien model [7]. The stationary filtered white noise models catch only the main frequency of the seismic waves that excite the structure and the bandwidth of the stochastic process. The stationary stochastic models generate artificial accelograms with constant amplitude, while the amplitude of the real accelograms is time-varying. To overcome this limit, several scholars have proposed non-stationary filtered white noise models to simulate the seismic ground motions. This kind of earthquake models is obtained from the product of a filtered stationary White Noise process and an envelope function dependent on the time (figure 1). In literature there are several different envelope functions: the research of a reliable envelope function to model the ground motion intensity has been the goal of many studies. Some of these functions are simple and deterministic, like the one proposed by Bolotin [1], others are complex. Jangid [8] has given an overview of different envelope functions. The main feature that distinguishes the envelope functions proposed in literature is its shape: it describes the temporal evolution of the amplitude of the ground shake (trapezoidal, double exponential, log-normal, etc.). The envelope functions have simple parametric forms and the values of the parameters are estimated from some characteristics of the earthquake records available for a specific area, like the duration of the strong ground motion, the energy of the seismic event and the kind of soil. Some studies have proposed envelope functions correlated with seismological parameters [9, 10, 11]. Unfortunately, these parameters are not significant and useful for the structural design. Previously Baker [12] has proposed a correlation of the ground motion intensity parameters used to predict the structural and geotechnical response.

In order to reproduce the temporal variation of the frequencies of the seismic input shaking the structure, evolutionary non-stationary stochastic model have been proposed. In these complex models the parameters of the stationary filtered stochastic process have a temporal variation. The temporal evolution of the frequency content of the accelograms is due to the different velocity of the P waves, S waves and surface waves that are released in the epicentre of the earthquake [13].

This study presented in this chapter proposes a new simple and effective deterministic envelope function that correlates the temporal variation of the amplitude of the seismic records to the most significant seismological parameters of the ones used in structural design: the PGA and the kind of soil. The shape of the proposed envelope function is based on the Saragoni and Hart's (SH) exponential function [14] with three parameters determined through an energetic criterion. This shape of the envelope function gives a very good agreement with the selected time-histories, as a numerical analysis shows hereafter. The proposed envelope function is calculated through a new procedure composed by two stages. In the first stage a deterministic pre-envelope mean function that is the real envelope of a set of selected earthquake records is estimated. The values of the two parameters of the envelope function for each selected accelerograms are estimated through an identification procedure. In the second stage a regression law for each parameter is estimated to generalize the results and to obtain values of the parameters of the envelope function useful in the seismic engineering.

The identification procedure of the parameters of the Modified Saragoni and Hart's function proposed here is based on the continuous energy release of the earthquake measured through the Arias Intensity (AI).

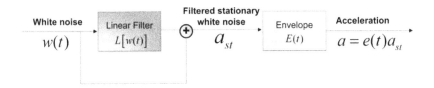

Figure 1. Scheme to simulate a stochastic ground motion

2. Stationary filtered stochastic process modelling earthquakes

As said in the introduction, a filtered stationary White Noise (WN) process $w(t)$ is the simplest model of the ones proposed to represent the stochastic seismic acceleration process. In this kind of models the WN process models the acceleration at the bedrock. One or more filters model the action of the soil between the epicentre and the basement of the structure. Indeed the soil filters the seismic waves and the resulting filtered signal has the mean frequency content of the earthquake acceleration recorded on the ground surface. The most famous model of filtered WN process that defines the ground motion acceleration is the Kanai-Tajimi (K-T) model. In this model the filtering effect of the soil is defined by a SDoF system, characterized by two parameters: the damping ratio ξ_g and the circular frequency ω_g. The ground acceleration a_{st} exiting the structure is the absolute acceleration of the filter. Therefore, the differential equations of the K-T model are:

$$\ddot{x} + 2\xi_g\omega_g\dot{x} + \omega_g^2 x = -w(t) \tag{1}$$

$$a_{st} = \ddot{x} + w(t) = -\left(2\xi_g\omega_g\dot{x} + \omega_g^2 x\right). \tag{2}$$

This filter is a linear second order one, so the Power Spectral Density (PSD) function of the filtered WN is

$$S(\omega) = S_0 \frac{\left[1 + \xi_g^2(\omega/\omega_g)^2\right]}{\left[1 - (\omega/\omega_g)^2\right]^2 + 4\xi_g^2(\omega/\omega_g)^2} \tag{3}$$

where S_0 is the PSD of the WN process [4, 5, 15]. This model has some limits: the amplitude of the ground acceleration is constant and the frequency content is constant in time. The K-T model fails in simulating earthquake ground motion characterized by medium and long duration, because it is described by only two filter parameters, the damping ratio and the circular frequency. In order to overcome this limit, Clough and Penzien [7] have proposed a model characterized by a double filtered WN process. The two filters that define the effect of the soil between the bedrock and the surface are linear. The differential equations of the C-P model are:

$$\begin{cases} a_{st}(t) = \ddot{x}_p(t) = -\omega_p^2 x_p(t) - 2\xi_p \omega_p \dot{x}_p(t) + \omega_f^2 x_f + 2\xi_f \omega_f \dot{x}_f(t) \\ \ddot{x}_p(t) + \omega_p^2 x_p(t) + 2\xi_p \omega_p \dot{x}_p(t) = \omega_f^2 x_f + 2\xi_f \omega_f \dot{x}_f(t) \\ \ddot{x}_f(t) + 2\xi_f \omega_f \dot{x}_f(t) + \omega_f^2 x_f = w(t) \end{cases} \tag{4}$$

where $x_f(t)$ is the response of the first filter characterized by the circular frequency ω_f and the damping ratio ξ_f, $x_p(t)$ is the response of the second filter characterized by the circular frequency ω_p and the damping ratio ξ_f and $w(t)$ is the exciting WN process. In this model the stationary ground acceleration coincides with the acceleration of the second filter output: $a_{st}(t) = \ddot{x}_p(t)$. In literature more complex models characterized by Multi-Degree of Freedom (MDoF) filters have be proposed to model seismic ground acceleration in a specific area. These models are defined by a larger number of parameters than the K-T and C-P models, so they model the soil filtering effect of the seismic waves better than the K-T and C-P models. On the other hand, these models with MDoF filters have a higher computational effort than the one of the K-T and C-P models. Some scholars [13, 16] have proposed non-stationary filtered WN models to take into account the temporal variation of the intensity of the acceleration records of the real earthquakes. The envelope function proposed in this study can be applied to modulate the amplitude in time of the stationary filtered White Noise obtained from a C-P model. Indeed, that model has four parameters correlated with the soil diffusion effect and reaches a good compromise between accuracy and computational effort to define seismic ground motion.

3. Envelope function definition

A Stationary stochastic process $\ddot{x}_p$ estimated through the C-P model is multiplied by an envelope function $E(t)$ to obtain a non-stationary filtered WN process with modulated intensity:

$$a(t) = \ddot{x}_p(t) E(t) \tag{5}$$

Each sample function of this non-stationary stochastic process is a synthetic accelerograms. The envelope function $E(t)$ proposed here is achieved from a complex procedure in two stages. A pre-envelope function defined from a set of selected recorded real accelerograms is estimated in the first step of the procedure. This pre-envelope function has the shape similar to the one developed by Saragoni and Hart [14] that is the most suitable for the set of the real records used. Indeed, this exponential deterministic envelope function has been chosen because it simulates better the strong ground motion than the other ones. It is continuous while other ones imply an arbitrary division into segments and the abrupt change in the frequency content of each segment. The Saragoni and Hart's (SH) function is

$$\varphi(t) = \alpha t^\eta e^{-\beta t} \qquad \alpha, \ \beta, \ \eta > 0, \tag{6}$$

where α, β and η are three calibration parameters. In order to reduce the number of the parameters of the SH function from three to two, in this study the parameters α and β are expressed as functions of the unknown time t_m in which the SH function has its maximum value. The maximum value of the modulation function is estimated from the system of equations:

$$\begin{cases} \varphi(t_m) = 1 \\ \dot{\varphi}(t_m) = 0 \end{cases} \tag{7}$$

From the solution of the eq. (7) the parameters α and β are valuated as function of the other two parameters η and t_m:

$$\beta = \frac{\eta}{t_m} \tag{8}$$

$$\alpha = \left(\frac{e}{t_m}\right)^\eta \tag{9}$$

Replacing the parameters α and β in the SH function (6) with the expressions (8) and (9), the SH function becomes

$$\varphi(\tau, \eta) = \tau^\eta e^{\eta(1-\tau)} \tag{10}$$

where $\eta = t / t_m$ and the independent variables are only η and t_m. In the follow this expression will be called Modified Saragoni and Hart (MSH) function. The intensity modulation of the earthquake ground motion is described not only by their peak value (Peak Ground Acceleration PGA), but also by the energy content or other quantities related to the energy content, such as the Arias Intensity (AI). The AI is defined as:

$$I_a = \frac{\pi}{2g} \int_0^{t_f} \ddot{x}_g^2(t) dt \tag{11}$$

and its mean value in stochastic terms is

$$\mu[I_a] = \frac{\pi}{2g} \int_0^{t_f} \langle \ddot{x}_g^2(t) \rangle dt = \frac{\pi}{2g} \sigma_{\ddot{x}_g}^2 \psi_a(t_f) \tag{12}$$

where t_f is total duration of the earthquake, $\ddot{x}_g(t)$ is the ground acceleration at time t in one of the two horizontal directions, g is acceleration due to gravity and the term $\psi_a(t_f)$ is defined by the expression

$$\psi(a) = \int_0^t \varphi^2(\rho) d\rho. \tag{13}$$

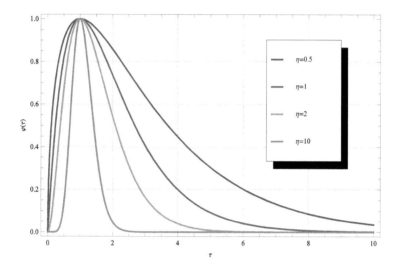

Figure 2. Sensitivity of the MSH function to the parameter η

Replacing the envelope function (10) in the expression (13), that expression becomes

$$\int_0^\tau \left[\varphi(\rho,\eta)\right]^2 d\rho = \tau^{2\eta}\left(\left(\frac{e}{2}\right)^{2\eta}(\eta\tau)^{-2\eta}\Gamma(2\eta)-\eta^{2\eta}\tau E_{-2\eta}(2\eta\tau)\right) \tag{14}$$

where $E_n(z)$ is the generalised exponential integral function

$$E_n(z)=\int_1^\infty \frac{e^{-z}}{n}d=z^{n-1}\Gamma(1-n,\,z), \tag{15}$$

while $\Gamma(z)$ is the Euler Gamma function

$$\Gamma(z)=\int_0^\infty z^{-1}e^{-d}. \tag{16}$$

The dimensionless ratio $\psi_a(\tau,\,\eta)/\psi_a(\tau_f,\,\eta)$ describes the energy release during earthquake event and it is plotted in the figure 3 for different value of the parameter η. This ratio is a function of the parameters τ and η. The parameter η determines the velocity of the energy release during a seismic event, while the parameter τ describes the energy release during the duration of the event with respect to the time of maximum amplitude of the accelerograms. The values of the parameters τ and η are estimated through an identification procedure.

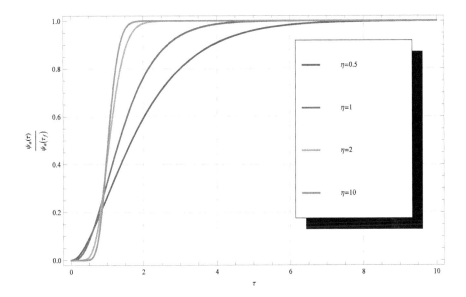

Figure 3. Sensitivity of the dimensionless ratio $\frac{\psi_a(\tau,\ \eta)}{\psi_a(\tau_f,\ \eta)}$ to the parameter η

4. Numerical procedure to evaluate the parameters of the pre-envelope functions

This section presents the identification procedure used to estimate the values of the parameters η and t_m that better characterize each of the selected real accelerograms of the PEER Next Generation Attenuation database.

The ground motion records of the PEER Next Generation Attenuation database that have been used in this study match the following criterion: the site where the seismic event is recorded has an average shear wave velocity in the top 30 meters comprised in four ranges according to the EC8 (B-C classes) and the NERPH classification (C-D classes) corresponding to stiff and soft soil respectively. The ground motion records of the PEER Next Generation Attenuation database are more than 7000 and half of them match this criterion. Further, the ground acceleration records of both the horizontal directions are used: for each selected accelogram of the database the weighted squared root of the sum of the squared east–west and north–south components is calculated and after it is used to estimate the pre-envelope mean function and the PGA used in the procedure proposed here.

The values of parameters η and t_m of the pre-envelope function are obtained by minimizing the difference of the ratios of the mean AI and the ratio of the term ψ_a describing the energy release estimated for analytical expression and real earthquake record:

$$\frac{\mu\left[I_a(t)\right]}{\mu\left[I_a(T_f)\right]} - \frac{\psi_a(t,\,t_m,\,\eta)}{\psi_a(T_f,\,t_m,\,\eta)}. \tag{17}$$

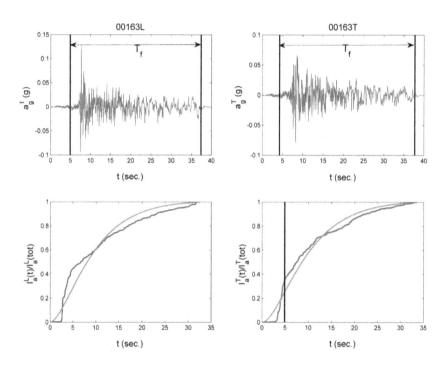

Figure 4. Record 00163 (kind of soil D) of the database and its AI. The blue lines indicates the L component and the red lines indicates the T component of the real record of the seismic event and the green lines indicates the analytical Areas Intensity

The ratios of the equation to minimize are functions of the total duration time of the earthquake accelerogram T_f. The total duration time is defined as the time during which the 98% of the total energy of the seismic signal is released. It is evaluated as the time interval between the 1% and 99 % of the AI of the seismic record. The total duration time is calculated for each selected earthquake record of the database. The values of the parameters η and t_m that describe the pre-envelope function for each real seismic event selected from the database seismic are achieved by means of an identification procedure formulated as an optimization problem. The Objective Function (OF) to minimize is

$$OF\left(\overline{b}\right) = \frac{1}{T_f}\int_0^{T_f} \left(\frac{I_a(\tau)}{I_a(T_f)} - \frac{\psi_a\left(\tau, \overline{b}\right)}{\psi_a\left(T_f, \overline{b}\right)}\right)^2 d\tau, \tag{18}$$

where $b = [t_m, \eta]$ is the vector of the variables. The optimization problem is solved by means of the Genetic Algorithm (GA) implemented in Matlab. To check the quality of this identification procedure of the parameters η and t_m for the pre-envelope function estimated for each earthquake record in the figures 4 and 5 the authors of this study have plotted a comparison between the AI of a real earthquake record and the analytical AI calculated using the parameters η and t_m estimated through identification procedure proposed. These two figures (4 and 5) show the comparison for two different earthquake records of the selected ones of the PEER Next Generation Attenuation database.

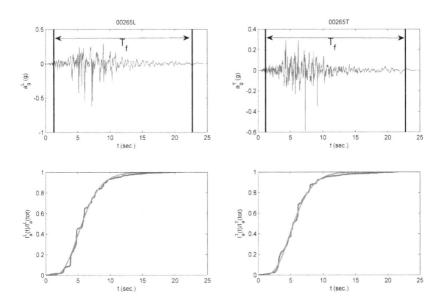

Figure 5. Record 00225 (kind of soil D) of the database and its AI. The blue lines indicates the L component and the red lines indicates the T component of the real record of the seismic event and the green lines indicates the analytical Areas Intensity

The identification of the parameters of the pre-envelope function is applied to each selected earthquake record of the PEER Next Generation Attenuation database. The selected earthquakes records of the PEER Next Generation Attenuation database and their identified parameters t_m and η are grouped according to four types of soil, as afore said. After the identification of the parameters η and t_m the mean AI is calculated:

$$\mu[I_a(t)] = \frac{\pi}{2g}\sigma^2_{\ddot{x}^{st}_g}\psi_a(t, t_m, \eta). \tag{19}$$

A linear function is used to correlate the PGA with its stationary variance $\sigma_{\ddot{x}^{st}_g}$ at the time of maximum amplitude of the accelerogram t_m:

$$PGA = \kappa\sigma_{\ddot{x}^{st}_g} \tag{20}$$

where

$$\kappa = PGA\sqrt{\frac{\pi}{2g}\frac{\psi_a(t_m)}{\mu[I_a(t_m)]}}. \tag{21}$$

The definition of the PGA is

$$PGA = max(\ddot{x}_g(\tau)|\tau[0, T_f]). \tag{22}$$

where T_f is the total duration of the record. The normalized value of the PGA is

$$max(\ddot{x}^N_g) = 1, \tag{23}$$

From the evaluation of (19), a new intensity measure I_e called Envelope Intensity (EI) is introduced:

$$I_e(T_f) = \frac{PGA}{g}\int_0^{T_f}a^N_g(\tau)d\tau. \tag{24}$$

5. Regression laws

In the second stage of the procedure to evaluate the envelope function described by the PGA and the kind of soil the regression laws that relate the parameters of the proposed envelope function with the PGA are extracted. The parameters of the envelope function to be identified are the total duration time T_f, the AI I_a, the maximum envelope time t_m, η and κ. The regression laws of the envelope function obtained from the pre-envelope functions of the seismic event selected from the database are

$$T_f(PGA) = T_f^{(0)}\left(e^{\left(T_f^{(1)}PGA\right)} + e^{\left(T_f^{(2)}PGA\right)}\right) \tag{25}$$

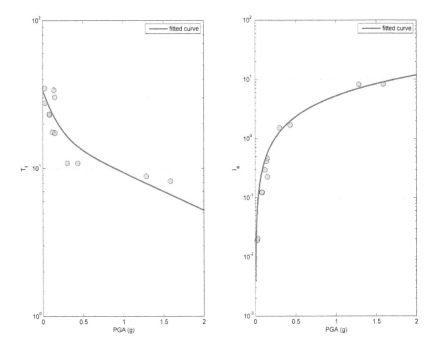

Figure 6. Values of I_a and T_f defined as function of the PGA. The value are related to the set of the seismic events selected from the database and occurred in sites characterized by kind of soil A.

$$I_a\left(PGA\right) = \exp\left(I_a^0 + I_a^1 \log\left(PGA\right)\right) \tag{26}$$

$$t_m\left(PGA\right) = t_m^0 e^{\left(t_m^1 PGA\right)} \tag{27}$$

$$\eta\left(PGA\right) = \eta^0 + \alpha_\eta PGA \tag{28}$$

$$\kappa\left(PGA\right) = \kappa^0 + \alpha_\kappa PGA. \tag{29}$$

In these equations the PGA is express in g (9.81 m/sec²). In the figure 6, 7, 8 and 9 it is fair that the curves of the regression law of the AI (eq. (26)) matches perfectly the trend of the AI valuated from the real data. The figures 6, 7, 8, 9, 10, 11, 12 and 13 show that the curves of the regression laws for the other parameters do not fit perfectly the numerical values of these parameters estimated for the selected accelograms of the database. The regression laws achieve one purpose of this study: the definition of analytical relations to estimate the most important parameters for different kinds of soil that characterize the amplitude modulation of earthquake records and the energy release of seismic events. The numerical values of these parameters for the four kinds of soil are collected in the table 1. These results can be used to calculate the envelope function that modulates the amplitude intensity of stationary filtered WN process to generate artificial accelerograms typical of a certain kind of the soil.

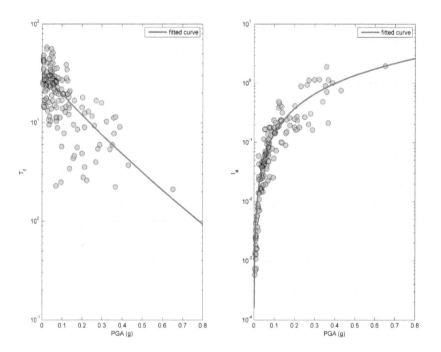

Figure 7. Values of I_a and T_f defined as function of the PGA. The value are related to the set of the seismic events selected from the database and occurred in sites characterized by kind of soil B.

Parameter		Soil type A	Soil type B	Soil type C	Soil type D
T_f^0 95% confidence bounds	[sec]	16.73 (9.371, 24.09)	16.12 (14.03, 18.2)	9.7 (8.189, 11.21)	14.76 (12.13, 17.4)
T_f^1 95% confidence bounds	[sec]	-0.582 (-1.496, 0.332)	-3.671 (-10.21, -2.87)	-10.32 (-17.69, -2.957)	-37.36 (-66.02, -8.699)
T_f^2 95% confidence bounds	[sec]	-6.367 (-19.57, 6.831)	-6.724 (-19.15, 5.706)	-0.7106 (-1.316, -0.105)	-1.057 (-1.841, -0.2726)
I_a^0 95% confidence bounds		1.668 (1.556, 1.78)	1.28 (1.092, 1.469)	1.543 (1.467, 1.618)	1.463 (1.351, 1.575)
I_a^1 95% confidence bounds		1.164 (0.9277, 1.4)	1.407 (1.247, 1.567)	1.568 (1.418, 1.717)	1.424 (1.312, 1.537)
t_m^0	[sec]	11.37	15.56	5.379	6.412
t_m^1	[sec]	-1.281	-6.58	-1.511	-2.185
η^0		1.798	3.4483	1.8058	1.8473
a_η		0.874	3.4498	4.4542	1.9887
κ^0		1.093	1.0446	1.6106	1.5914
a_κ		0.652	2.0301	0.6132	0.5678

Table 1. Parameters values of the regression laws estimated for all the four kinds of soil.

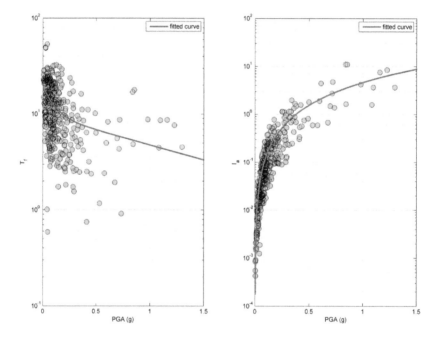

Figure 8. Values of I_a and T_f defined as function of the PGA. The value are related to the set of the seismic events selected from the database and occurred in sites characterized by kind of soil C

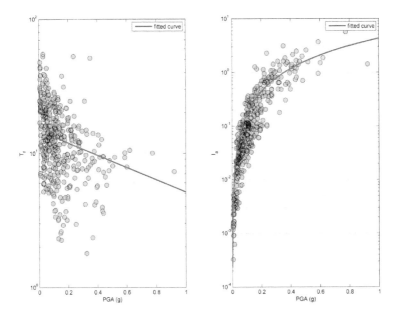

Figure 9. Values of I_a and T_f defined as function of the PGA. The value are related to the set of the seismic events selected from the database and occurred in sites characterized by kind of soil D.

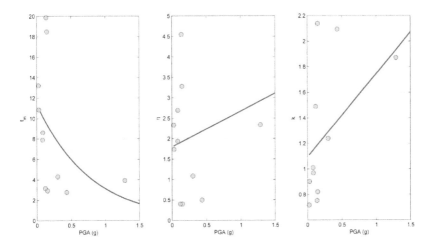

Figure 10. Values of t_m, η and κ defined as function of the PGA. The value are related to the set of the seismic events selected from the database and occurred in sites characterized by kind of soil A.

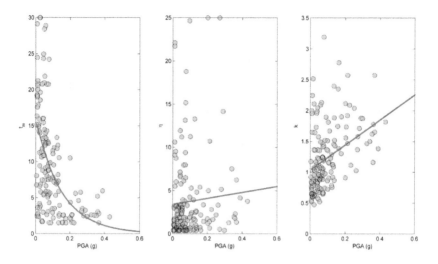

Figure 11. Values of t_m, η and κ defined as function of the PGA. The value are related to the set of the seismic events selected from the database and occurred in sites characterized by kind of soil B.

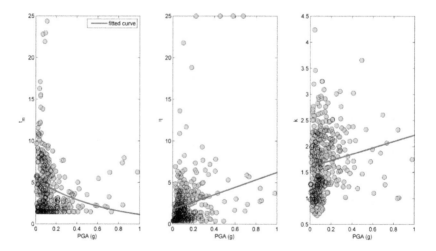

Figure 12. Values of t_m, η and κ defined as function of the PGA. The value are related to the set of the seismic events selected from the database and occurred in sites characterized by kind of soil C.

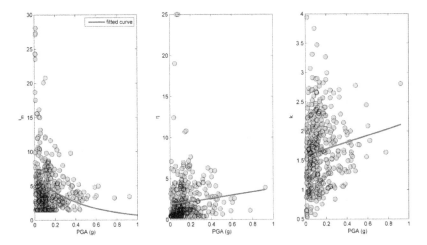

Figure 13. Values of t_m, η and κ defined as function of the PGA. The value are related to the set of the seismic events selected from the database and occurred in sites characterized by kind of soil D.

6. Conclusions

One of the main problems of earthquake engineering is the proper estimation of the characteristics of future earthquakes that will affect new and existing structures. This is a non-trivial problem because of the inner unpredictable nature of earthquakes. Due to this nature of earthquakes, stochastic models have been proposed to generate synthetic future seismic accelerograms to use in the structural design. Some of the stochastic models already proposed relate the stochastic ground motion process to seismological parameters that are not meaningful in structural engineering. The study here proposed overcomes this limit: it presents a model that describes the earthquake ground motion in term of parameters useful in the structural engineering. This model is a non-stationary stochastic one based on the stationary C-P model and characterized by the temporal modulation of the amplitude. The amplitude modulation is produced by a new envelope function that has the same shape of the SH function, but it is described by only two parameters. In order to obtain the values of these parameters of the envelope function a complex procedure is used. The procedure has two stages:

1. The estimation of parameters for each of selected accelerograms of the PEER Next Generation Attenuation database to generate a pre-envelope function of each accelogram.

2. The regression analysis of the values of these parameters to obtain their mean values for a class of soil.

In order to estimate the values of the parameters of the envelope function for a class of soil other analytical results are obtained. Relations of the parameters of the MSH envelope function with other characteristics of the earthquake ground motions and the AI are imposed (eqs. (13), (19), (20) and (21)), so analytical formulae to estimate other characteristics of the seismic events in term of PGA and kind of soil are obtained from the regression analysis (eqs. (25), (26), (27), (28), (29)).

The envelope function here presented and the method to calculate its parameters produce a temporal modulation of the amplitude in the synthetic accelograms in term of the most significant quantities used in structural engineering: the PGA and the kind of soil.

Finally, the numerical values of the characteristics of the earthquake ground motion obtained from the regression analysis are collected in table to be used in future applications of the earthquake engineering.

Author details

Giuseppe Carlo Marano[1], Mariantonietta Morga[2] and Sara Sgobba[1]

1 Department of Civil and Architectural Science, Technical University of Bari, Bari, Italy

2 Mobility Department – Transportation Infrastructure Technologies, Austrian Institute of Technologies GmbH, Vienna, Austria

References

[1] Bolotin VV. Statistical theory of the aseismic design of structures. In: 2nd World Conference on Earthquake Engineering - Tokio, volume 2, pages 1365-1374. Science Council of Japan, 1960.

[2] Housner GW and Jennings PC. Generation of artificial earthquakes. ASCE, Journal of the Engineering Mechanics Division, 1964; 90: 113–150.

[3] Housner GW. Characteristics of strong-motion earthquakes. Bulletin of the Seismological Society of America, 1947; 37(1) 19–31.

[4] Kanai K. Semi-empirical formula for the seismic characteristics of the ground motion. Bulletin of the Earthquake Research Institute, 1957; 35: 309–325.

[5] Tajimi H. A statistical method of determining the maximum response of a building structure during an earthquake. In: 2nd World Conference on Earthquake Engineering - Tokio, volume 2, pages 781–798. Science Council of Japan, 1960.

[6] Sgobba S, Marano GC, Stafford PJ, and Greco R. New Trends in Seismic Design of Structures, chapter: Seismologically consistent stochastic spectra. Saxe-Coburg Publisher, 2009.

[7] Clough RW and Penzien J. Dynamics of structures, Mc Graw Hill; 1975.

[8] Jangid RS. Response of SDoF system to non-stationary earthquake excitation. Earthquake Engineering & Structural Dynamics, 2004; 33(15) 1417–1428.

[9] Campbell KW. Prediction of strong ground motion using the hybrid empirical method: example application to eastern-north America. Bulletin of the Seismological Society of America, 2002; 93(3) 1012–1033.

[10] Cua G. Creating the Virtual Seismologist: Developments in Ground Motion Characterization and Seismic Early Warning. PhD thesis, California Institute of Technology, 2005.

[11] Stafford PJ, Sgobba S, and Marano GC. An energy-based envelope function for the stochastic simulation of earthquake accelerograms. Soil Dynamics and Earthquake Engineering, 2009; 29(7) 1123–1133.

[12] Baker JW. Correlation of ground motion intensity parameters used for predicting structural and geotechnical response. In: ICASP10 - 10th International Conference on Applications of Statistics and Probability in Civil Engineering, Tokyo, Japan, 2007.

[13] Conte JP and Peng BF. Fully nonstationary analytical earthquake ground-motion model. ASCE, Journal of Engineering Mechanics, 1997; 12: 15–24.

[14] Saragoni GR and Hart GC. Simulation of artificial earthquake. Earthquake Engineering and Structural Dynamics, 1974; 2: 249–267.

[15] Marano GC, Morga M and Sgobba S. Modelling of stochastic process for earthquake representation as alternative way for structural seismic analysis: past, present and future. In: EQADS 2011 - International Conference on Earthquake Analysis and Design of Structures - Department of Civil Engineering, PSG College of Technology, Coimbatore, Tamilnadu, India, December 1-3 2011.

[16] Amin M and Ang AHS. Nonstationary stochastic model of earthquake motions. ASCE, Journal of the Engineering Mechanics Division, 1968; 94: 559–583.

Chapter 7

Characterizing the Noise for Seismic Arrays: Case of Study for the Alice Springs ARray (ASAR)

Sebastiano D'Amico

Additional information is available at the end of the chapter

1. Introduction

A seismic array is defined as a suite of seismometers with similar characteristics. Seismic array were originally built to detect and identify nuclear explosions. Since their development all over the world, seismic arrays have contributed to study interior of volcanoes, continental crust and lithosphere, determination of core-mantle boundary and the structure of inner core. Seismic arrays have been used to perform many regional tomographic studies (e.g., Achauer and the KRISP Working Group, 1994; Ritter et al., 1998, 2001); they helped to resolve fine-scale structure well below the resolution level of global seismology in many different places in the Earth, from the crust using body waves (e.g., Rothert and Ritter, 2001) and surface waves (e.g., Pavlis and Mahdi, 1996; Cotte et al., 2000), the upper mantle (e.g., Rost and Weber, 2001), the lower mantle (e.g., Castle and Creager, 1999), the core-mantle boundary (e.g., Thomas et al., 1999; Rost and Revenaugh, 2001), and the inner core (e.g., Vidale et al., 2000; Vidale and Earle, 2000; Helffrich et al., 2002). A different branch of seismology that benefited from arrays is "forensic seismology" (Koper et al., 1999; 2001; Koper and Wallace 2003). Studied have been also carried out to track the rapture propagation of large and moderate earthquakes (Goldstein and Archuleta 1991a,b: Spudich and Cranswick 1984; Huang 2001; D'Amico et al. 2010; Sufri et al. 2012; Koper et al. 2012), studies related to the seismic noise have been also developed using seismic arrays (Koper and Fathei, 2007; Gerstoft et al. 2006; D'Amico et al. 2008; Schulte-Pelkum et al., 2004). For example Gerstoft et al. (2006) used beamforming of seismic noise recorded on California Seismic Network to identify body and surface waves generated by the Hurricane Katrina. Schulte-Pelkum et al., (2004) measured direction and amplitude of ocean-generated seismic noise in the western United States. Koper and Fatehi (2007) used 950, randomly chose, 4-sec long time windows from 1996 to 2004 at the CMAR array located in Thailand. In their work they found, around 1Hz, a large noise peak coming from southwest near 220 degrees and an apparent velocity of 3.5-4.0km/s. Their results are robust from year-to-year and are also consistent from season to season. Two lesser noise peaks show probably a seasonal dependence, being much stronger in the fall and winter than in the summer and spring. Neither peak is sensitive to the "hour-to-hour" analysis meaning they are uncorrelated to anthropical noise. Koper and De Foy (2008) showed that the seismic noise recorded at the CMAR array during 1995-2004 can be strongly correlated with the ocean wave's heights. They carried out this information by using data from TOPEX/POSEIDON satellite tracks and explained them by the local monsoon-driven climate. For all this different purpose a lot of different arrays techniques and methods have been developed (for reviews see: Rost and Thomas, 2002; Filson, 1975) and applied to a wide number of high-quality data set.

The main goal of this chapter is to highlight the main characteristics of noise for the Alice Springs ARray (Australia). Furthermore detecting the noise we would like, if it exists, try to found the large peak noise, the predominant direction and estimate the optimal phase velocity and eventual time dependence. This kind of study could play a key role in for the isolation of the seismic noise in designing new arrays or particular instruments such as the construction of gravitational wave detectors (Hoffmann et al.,2002 and reference in therein). Theoretically knowing the seismic noise features and source it will be possible to subtract its effect from the data.

2. Data set and processing

Alice Springs Array is located in Australia and it is made by 19 vertical component short period seismographs deployed with an effective aperture of about 10 km (Fig.1). We ignored elevation differences among the array elements and considered only 2D wavenumber vectors. This is a reasonable since the ASAR array is relatively flat.

Continuous data were available from 1994 to 2004 and it was possible to get them by using the "autodrm request" of the U.S. Army Space and missile Defense Command monitoring research program (www.rdss.info, last access in 2009). It supports different researches related to the nuclear explosions monitoring. Time series containing randomly noise recorded for each station in all the time period where recordings are available. In the present paper data from 1999 to 2001 are used. We extracted several minutes of continuous data once a week for the selected time period, making sure to vary the time of day and the day of the week (Fig. 2). We used the Generic Array Processing software (GAP; Koper 2005), a set of freely C programs for processing seismic array data. These programs operate on binary SAC files and output GMT (Wessel and Smith, 1991) scripts for visualizing the results; they were developed to work both with small aperture array and other type of array as well. In present paper we used in particular the program called "capon.c" that performs the signal processing following the maximum likelihood Capon (1969) method; the idea is to use a spectral density function that provides the information concerning the power as a function of frequency, this function also provides the vector velocities of the propagating waves. This kind of approach is also known in literature as frequency-wavenumber (f-k) analysis; it offers the opportunity to determine the back-azimuth and the slowness of coherent seismic waves with a high resolution. Furthermore, it has the possibility to detect and discriminate simultaneously several microseismic sources. Each trace was examined to eliminate those with spurious transients or glitches, null traces and those contain obvious earthquake energy. After this selection the original dataset was reduced about of the 5%; each time window is 5 minutes long. Figure 3 shows a schematic diagram of the method applied in this study. The analyses are performed at different frequency bands (around 0.4Hz, 0.6Hz, 0.8 Hz, 1.0Hz, 2.0Hz and 4.0Hz).

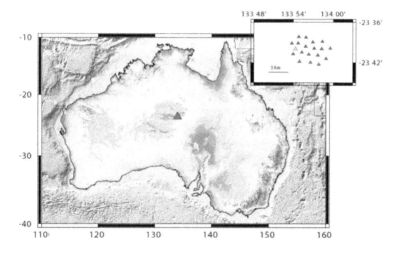

Figure 1. Alice Springs Array (ASAR) location and array geometry. The white triangle in the top panel represents the reference element, while the dark gray triangles are the other 18 elements of the array

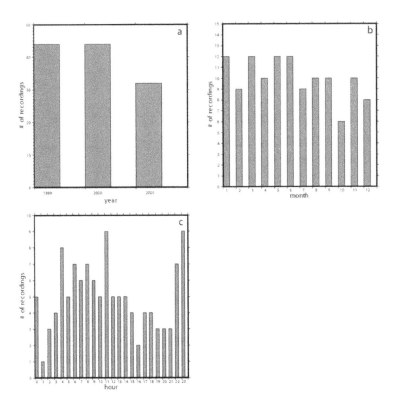

Figure 2. Characteristics of our data set of seismic noise recorded by ASAR. (a) number of recording as function of year, month (b) and hours (UTC) (c)

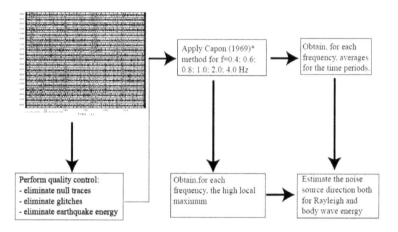

Figure 3. Example of used time windows and schematic representation of the procedure applied in the present study

3. Results and discussion

Figures 4 shows the number of recording as a function of the optimal ray parameter for the frequencies of 0.4, 0.6, 0.8, and 1.0 Hz. Figures 5, 6, 7, 8, 9, 11, and 12 show the time-averaged ambient noise field at ASAR array averaged per each year and the average for the three-year period and the three-year period respectively, binned according to month at the frequency of 0.4Hz, 0.6Hz, 0.8 Hz, 1.0Hz. Figure 13 reports the results obtained for 2.0 and 4.0Hz respectively and showing the average per each year and the average for the three-year period. Figure 14 plots the local maxima (from 0.4 to 1.0 Hz) computed using the Capon (1969) method; red dots represent all the local maxima while blue are the maxima having a relative power greater than 5db. We observed, for the frequencies of 0.4, 0.6, 0.8 and 1.0 Hz, the most prominent pick coming from the S-W direction with an optimal backazimuth around 190-200 degrees and an apparent velocity of about 3-4km/s indicative of higher mode Rayleigh waves. This energy is probably generated as waves from the interaction of oceanic waves with the coast in the Australian Bight. Because of the high attenuation of short period Rayleigh waves, it is really unlikely that the noise is generated further away from the ASAR array. It is also possible to highlight a possible correlation between noise peaks and the distance of the array to the coast line. In fact, according the plot in figure 14 for each different frequency it is possible to notice that the largest number of peak having a relative power greater than 5db is coming from the S-W direction; the second large number of peak is coming from the N-E direction and a very few are coming from the S-E part that is the largest distance from the coast. An other important noise peak shown in figure 14 occurs in the center of the plot, indicating that energy is coming almost with a vertical incidence on the array. There is not any peak for the high-frequencies (f=2.0; 4.0 Hz); that is probably due to the location of the array. Furthermore we can also point our attention on the amplitude as a function of time (fig. 15). It seems there are some seasonal patterns, in fact, the maximum peaks occur in the winter time while the minimum values are during the summer time (please remember that the array is located in the Southern Hemisphere).

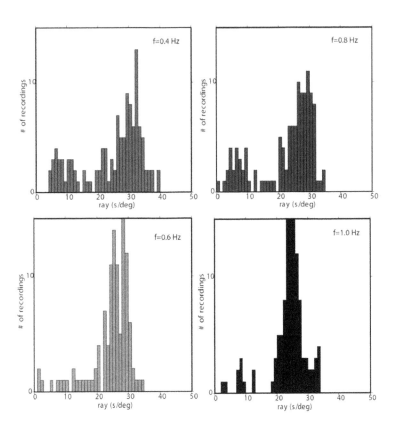

Figure 4. Number of recordings as a function of the optimal ray parameters for different frequency

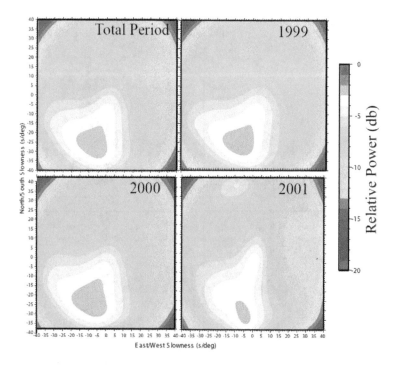

Figure 5. Average of the relative power per year at 0.4 Hz

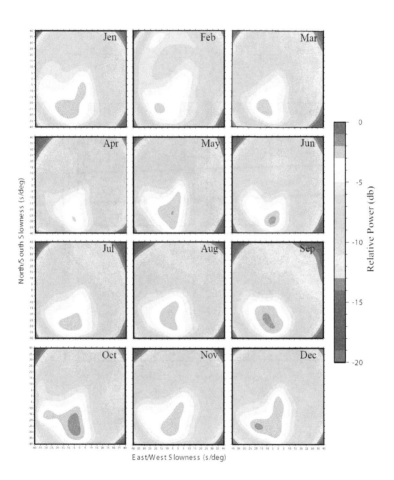

Figure 6. Average of the relative power per month at 0.4 Hz

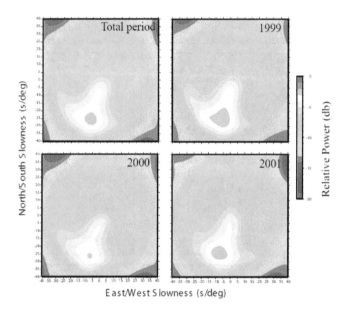

Figure 7. Average of the relative power per year at 0.6 Hz

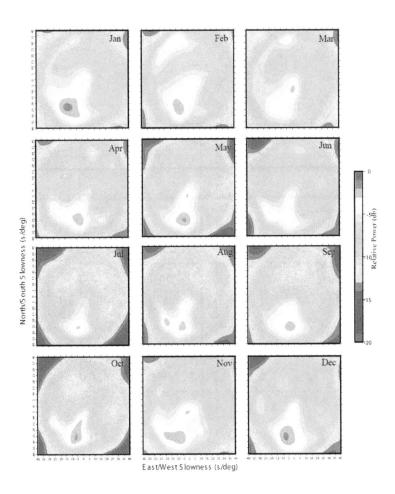

Figure 8. Average of the relative power per month at 0.6 Hz

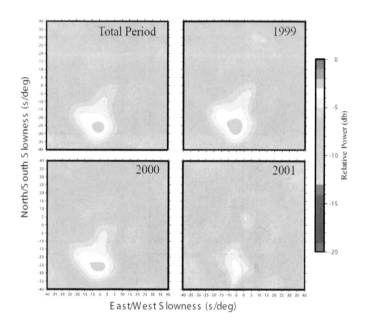

Figure 9. Average of the relative power per year at 0.8 Hz

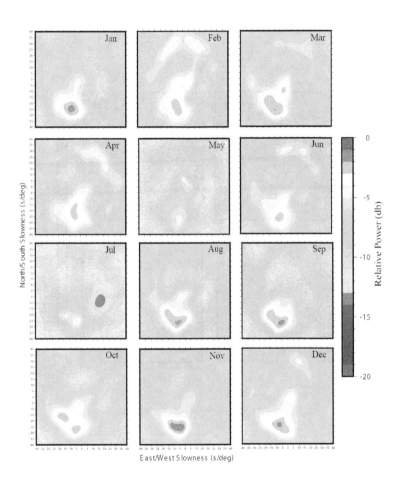

Figure 10. Average of the relative power per month at 0.8 Hz

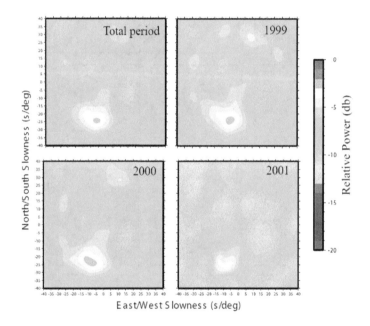

Figure 11. Average of the relative power per year at 1 Hz

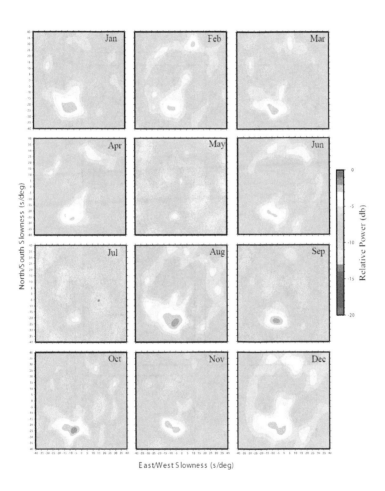

Figure 12. Average of the relative power per month at 1 H

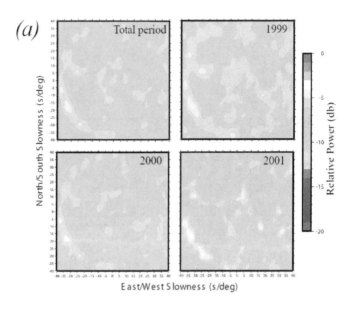

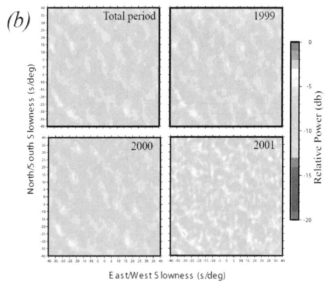

Figure 13. Average of the relative power per year at 2 Hz (a) and 4 Hz (b)

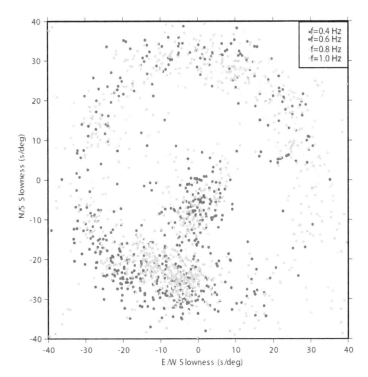

Figure 14. Local maxima computed using the Capon (1969) method having a relative power greater than 5db for different frequencies It is possible to notice a relationship between the maxima in the S-W and N-E direction; they seem be quite spread in the N-W and S-E directions; perhaps due to the distance between the array and the coast lines.

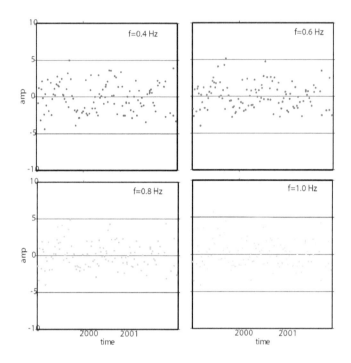

Figure 15. Maximum amplitude versus time

4. Concluding remarks

Seismic array have contributed to develop different studies to investigate the interior of the Earth. In this paper we used some array techniques in order to highlight the charac-teristic of noise for a relative small aperture array (about 10 km): the ASAR array locat-ed in central Australia. We used waveforms from 1999 to 2001 choosing the data in order to cover each year, month day and part of it. We used the Capon (1969) method and we performed the analysis at different frequencies (0.4, 0.6, 0.8, 1.0, 2.0 and 4.0 Hz). For each frequency the optimal ray parameter, the optimal phase velocity and the opti-mal backazimuth were calculated. Results show that there is a consistent peak for the optimal backazimuth around 190-200 degrees for the frequency ranged from 0.4 to 1.0 Hz; the maximum peak disappears for the 2.0 and 4.0Hz analysis. The predominant peak in the S-W direction could be interpreted as ocean waves interacting with the coast in the Australian Bight. The absence of peaks for the analysis above the 2.0 Hz confirm that there is no evidence of anthropical noise, that is probably due to the location of the array. We found a maximum peak around 3-4Km/s for the phase velocity indicative of higher-mode Rayleigh waves. Some dispersion is evident in the phase velocity peaks, and the large noise peak to the southwest is consistent from season to season, suggesting that there are some seasonal patterns as well. In some of the f-k spectra it is possible to notice a double peak, in which there appears to be a body-wave component to the noise.

The author thanks Dr. Keith Koper (University of Utah, USA) for providing the Generic Array Processing software. Data where obtained using the "autodrm request" of the U.S. Army Space and missile Defense Command monitoring research program (www.rdss.info, last access in 2009). The author is also very thankful to Ms. Silvia Vlase for her support.

Author details

Sebastiano D'Amico

Address all correspondence to: sebdamico@gmail.com

Depaertment of Physics, University of Malta, Msida, Malta

References

[1] Achauer, U., and the KRISP Working Group, 1994. New ideas on the Kenya rift based on the onversion of the combined dataset of the 1985 and 1989/90 seismic tomography experiments, *Tectonophysics,* 236, 305–329.

[2] Capon J., 1969. High resolution frequency-wavenumber spectrum analysis, Procceding of the IEEE, 57, 8, 1408-1418

[3] Castle, J. C., and K. C. Creager, 1999. A steeply dipping discontinuity in the lower mantle beneath Izu-Bonin, *J. Geophys. Res.,* 104, 7279–7292.

[4] Cotte, N., H. A. Pedersen, M. Campillo, V. Farra, and Y. Cansi, 2000. Of great-circle propagation of intermediate-period surface waves observed on a dense array in the French alps, *Geophys. J. Int.,* 142, 825–840.

[5] D'Amico S., Koper K.. D., Herrmann R B., 2008. Array analysis of short-period seismic noise recorded in central Australia. *Seismological Research Letters,* vol. 79, 2, 293

[6] D'Amico S., Koper K. D., Herrmann R. B., Akinci A., Malagnini L., 2010. Imaging the rupture of the M_w 6.3 April 6, 2009 L'Aquila, Italy earthquake using back-projection of teleseismic P-waves. *Geophysical Research Letters,* 37, L03301, doi: 10.1029/2009GL042156

[7] Filson J., 1975. Array seismology. *Annual Reviews, earth Planet. Sci.,* 3 157-181.

[8] Gerstoft P.M., Fehler M.C., Sabra K.G., 2006. When Katrina hit California; *Geophys Res. Lett.*, 33, L17308.

[9] Goldstein and Archuleta, 1991a. Deterministic Frequency Wavenumber methods and direct measurements of rapture propagation during earthquakes using a dense array: theory and methods. *Journal of Geophysical Research*, 98, 6173-6185.

[10] Goldstein and Archuleta, 1991b. Deterministic Frequency Wavenumber methods and direct measurements of rapture propagation during earthquakes using a dense array: data analysis. *Journal of Geophysical Research*, 98, 6187-6198.

[11] Hoffmann, H., Winterflood, J., Cheng Y., Blair D. G., 2002. Cross-correlation studies with seismic noise, *Class. Qaunt. Grav.*, 19, 1709-1716

[12] Huang B-S., 2001. Evidence for azimuthal and temporal variations of the rapture propagation of the 1999 Chi-Chi, Taiwan earthquake from dense seismic array observations, *Geophys. Res. Lett.*, 28, 17 3377-3380.

[13] Helffrich, G., S. Kaneshima, and J.-M. Kendall, 2002. A local, crossing-path study of attenuation and anisotropy of the inner core, *Geophys. Res. Lett.*, 29, 1568.

[14] Koper K., 2005, The Generic Array Processing (GAP) Software Package, SSA meeting web: http://www.eas.slu.edu/People/KKoper/Free/index.html (last access Dec-17-2007)

[15] Koper K and Fathei A. (2007). Modeling P wave multipathing at regional distances in southeast Asia. *Final Scientific report # FA8718-06-C-003.*

[16] Koper, K.D., T.C. Wallace, S.R. Taylor, and H.E. Hartse, 2001. Forensic seismology and the sinking of the Kursk, *EOS Trans., AGU, 82*, pp. 37,45-46

[17] Koper, K.D., T.C. Wallace, and D. Hollnack, 1999. Seismic analysis of the 7 August 1998 truck-bomb blast at the American Embassy in Nairobi, Kenya, *Seismol. Res. Lett.,* 70, 512-521.

[18] Koper, K.D., T.C. Wallace, R.C. 2003. Aster, Seismic recordings of the Carlsbad, New Mexico pipeline explosion of 19 August 2000, *Bull. Seism. Soc. Am., 93*, 1427-1432

[19] Koper, K.D. and B. de Foy (2008), Seasonal anisotropy in short-period seismic noise recorded in South Asia, Bull. Seismol. Soc. Am., 98, 3033-3045.

[20] Koper, K. D., A. R. Hutko, T. Lay, and O. Sufri (2012), Imaging short-period seismic radiation from the 27 February 2010 Chile (Mw 8.8) earthquake by back-projection of P, PP, and PKIKP waves, J. Geophys. Res., 117, B02308, doi:10.1029/2011JB008576.

[21] Kraft T., Wassermann J., Schmedes E., Igel H., 2006. Metereological triggering of earthquake swarms at Mt. Hochstaufen, SE-Germany, *Tectonophysycs*, 424, 245-258.

[22] Pavlis, G. L., and H. Mahdi, 1996. Surface wave propagation in central Asia: Observations of scattering and multipathing with the Kyrgyzstan broadband array, *J. Geophys. Res.*, 101, 8437–8455.

[23] Ritter, J. R. R., U. R. Christensen, U. Achauer, K. Bahr, and M. Weber, 1998. Search for a mantle plume under central Europe, *Eos Trans. AGU*, 79, 420.

[24] Ritter, J. R. R., M. Jordan, U. Christensen, and U. Achauer, 2001.A mantle plume below the Eifel volcanic fields, Germany, *Earth Planet. Sci. Lett.*, 186, 7–14.

[25] Rost S., C. Thomas, 2002. Array seismology: methods and applications. *Reviews of Geophys.*, 40, 3

[26] Rost, S., and J. Revenaugh, 2001. Seismic detection of rigid zones at the top of the core, *Science*, 294, 1911–1914.

[27] Rost, S., and M. Weber, 2001. A reflector at 200 km depth beneath the NW Pacific, *Geophys. J. Int.*, 147, 12–28.

[28] Rothert, E., and J. R. R. Ritter, 2001. Small-scale heterogeneities below the Gra°fenberg array, Germany, from seismic wave-field fluctuations of Hindu Kush events, *Geophys. J. Int.*, 140, 175–184.

[29] Schulte-Pelkum V., Earle P.S., Vernon F.L., 2004. Strong directivity of ocean-generated seismic noise; *Geochem. Geophys. Geosyst.*, 5, Q03004.

[30] Spudich P., E. Cranswick, 1894. Direct observation of rapture propagation during the 1979 imperial valley earthquake using a short baseline accelerometer array. *Bull. Seism. Soc. Am.*, 74, 6, 2083-2114.

[31] Sufri, O., K. D. Koper, and T. Lay (2012), Along-dip seismic radiation segmentation during the 2007 Mw 8.0 Pisco, Peru earthquake, Geophys. Res. Lett., 39, L08311, doi: 10.1029/2012GL051316.

[32] Thomas, C., M. Weber, C. Wicks, and F. Scherbaum, 1999. Small scatterers in the lower mantle observed at German broadband arrays, *J. Geophys. Res.*, 104, 15,073–15,088.

[33] Vidale, J. E., and P. S. Earle, 2000. Fine-scale heterogeneity in the Earth's inner core, *Nature*, 404, 273–275.

[34] Vidale, J. E., D. A. Dodge, and P. S. Earle, 2000. Slow differential rotation of the Earth's inner core indicated by temporal changes in scattering, *Nature*, 405, 445–448.

[35] Wessel P. and Smith W. H. F., 1991. Free software helps map and display data, *Eos Trans.*, AGU, 72, 441

Damage Estimation Improvement of Electric Power Distribution Equipment Using Multiple Disaster Information

Yoshiharu Shumuta

Additional information is available at the end of the chapter

1. Introduction

Electric power distribution equipment has a high damage potential due to disasters and requires a long time for restoration. This is because a huge number of electric power distribution equipment is installed under various ground and regional conditions and is located near vulnerable trees and old residential buildings. Thus, during the restoration work after a large scale earthquake, it takes a long time to collect reliable disaster information. It is also even difficult to accurately estimate the damage degree of electric power distribution equipment. Therefore, in Japan, electric power companies pay particular attention to technologies associated with a quickly understanding and estimating the damage degree of the entire electric power distribution system during the emergency restorations.

On the other hand, with the progress in information technologies, practical disaster information services are increasing in Japan. For example, the Japan Metrological Agency has started a general delivery service of real-time earthquake information since October, 2007[1]. Moreover, in recent years, remote sensing images, such as satellite, aero and synthetic aperture radar (SAR) images, are now available and open to the public after a large scale disaster [2]. On the basis of such information technologies, more reasonable ways to support the restoration work for utility lifelines can be developed.

Based on this background, our research team has developed an sequentially updated damage estimation system called RAMPEr, which stands for "*Risk Assessment and Management System for Power lifeline Earthquake real time*" [3][4]. RAMPEr enables us to provide the damage estimation results of electric power distribution equipment during the emergency restoration process against an earthquake.

Reference [3] proposed a damage estimation function installed in RAMPEr which used the earthquake ground motion intensity as an input parameter, and applied the proposed function to actual electric power distribution equipment damaged by the 2007 Niigata-ken Chuetsu Oki earthquake to clarify the estimation accuracy of the proposed function.

This paper focuses on the updated damage estimation process of RAMPEr. RAMPEr enables us to improve the damage estimation accuracy using sequentially updated disaster information which a power company can collect during the emergency restoration period against a large-scale earthquake. Chapter 2 introduces the necessary disaster information for three emergency restoration stages and emphasizes the significance of RAMPEr. Chapter 3 describes the formulation of the proposed model. Chapter 4 discusses the advantage and limitation of the proposed model using the actual damage records due to the 2007 Niigata-Ken Chuetsu-Oki earthquake.

2. Information required for the emergency restoration work

Fig.1 shows the differences in the information required for the efficient restoration of a seismic damaged electric power distribution system. The restoration process is generally divided into the initial, emergency, and permanent restoration periods. During the initial restoration period, the information associated with the damage degree of the entire electric power system is needed to judge how many staff members should be dispatched for the restoration. During the emergency restoration period, the information associated with the damage point and mode to judge whether restoration staff members can immediately restore the power is needed. During the permanent restoration process, the information associated with all damaged equipment to be physically repaired is needed.

In order to collect this information, the power company dispatches inspection teams. In addition, some power companies have tried to apply remote sensing technologies, including helicopters and satellites, to quickly collect damage information. However, at the current time, the inspection teams and remote sensing technologies are not very effective to quickly collect seismic damage information especially for the restoration resource allocation during the initial restoration period.

Fig.2 shows the restoration process of an electric power distribution system located in the Tohoku region just after the 2011 earthquake off the Pacific coast of the Tohoku (the 311 earthquake) occurrence. The horizontal axis and the vertical axis show the elapsed time (days) and the number of inspected damaged equipment (%), respectively. Fig.2 indicates that the damage information had not been effectively collected, especially during the initial restoration period. This is because the Tohoku area had frequent aftershocks, much debris, and some coastal regions which were not able to be entered for the restoration analysis. This result illustrates that when a large-scale earthquake occurs, there is the possibility that the restoration work, including damage information collection, is highly limited by the damage related to residential buildings and other infrastructures.

During the limited damage information condition, RAMPEr becomes an effective tool to support some decision makings. RAMPEr is a seismic damage estimation system whose basic concept is a sequential updated damage estimation based on real-time hazard and damage information. RAMPEr was used by Tohoku Electric Power Co., Inc., to support its actual initial restoration work due to the 311 earthquake [10].

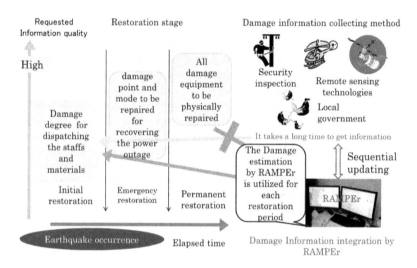

Figure 1. Differences in required information according to restoration stages

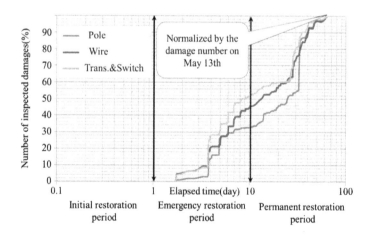

Figure 2. Time required for the damage detection of electric power distribution equipment [9]

3. Consecutive integrated process of multiple disaster information

3.1. Multiple disaster information used by RAMPEr

Disaster information needed by an electric power company can be usually collected after an earthquake occurrence includes four categories; (1) Earthquake, (2) Power outage, (3) Damage inspection, and (4) Damaged area image by remote sensing.

As for the earthquake information, RAMPEr, which has been already installed in some electric power companies, is supposed to automatically receive the earthquake information through the Internet including the epicenter, magnitude, and seismic ground motion intensities recorded on seismographs from the Japan Metrological Agency within several minutes just after the earthquake occurrence. Based on the received earthquake information, RAMPEr evaluates the seismic ground motion intensity distribution. As other sources, the National Research Institute for Earth Science and Disaster Prevention (NIED) opens seismic ground motion records including K-NET and KiK-net[5]. RAMPEr collects Instrumental Seismic Intensity (ISI), Peak Ground Velocity (PGV), and Peak Ground Acceleration (PGA) recorded from K-NET and KiK-net to improve the evaluation accuracy of the seismic ground motion distribution.

Power outages and damage inspection information are usually confidential information that the power company collects. The power outage information is usually collected for every high voltage distribution line (feeder) with power outages obtained from an online business support system maintained by the electric power company. The inspection information includes damaged equipment information including distribution poles, distribution lines, transformers, and switches. The inspection information is collected by portable transceivers, mobile phones and Personal Digital Assistances (PDA) as offline information. RAMPEr uses this confidential information to improve the damage estimation accuracy [4].

On the other hand, a seismic damage area image provided by remote sensing devices, such as satellites, is also one of the effective resources to understand the damage degree of the earthquake stricken area. However, as mentioned in Chapter 2, at the current technology stage, because it usually takes a long time to take and provide the images, it is difficult for RAMPEr to get the satellite image during an emergency restoration period. Thus, this paper focuses on (1) Earthquake information, (2) Power outage information, and (3) Damage inspection information to formulate the damage information integration as follows.

3.2. Basic idea of damage information integration

This paper focuses on a Bayesian network as a basic model to integrate the multiple disaster information. Details of the Bayesian network can be found in reference [6]. This chapter only introduces the basic concept of the Bayesian network for a better understanding of the following chapters.

Fig.3 shows a simple Bayesian network. It describes the relationships between variables X_1 and X_2. The X_1 and X_2 variables are binary (taking a value of either 0: false or 1: true). The

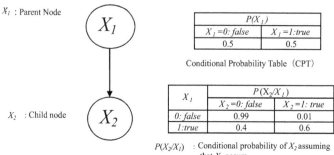

X_1 : Parent Node

$P(X_1)$	
$X_1=0$: false	$X_1=1$: true
0.5	0.5

Conditional Probability Table (CPT)

X_2 : Child node

X_1	$P(X_2/X_1)$	
	$X_2=0$: false	$X_2=1$: true
0: false	0.99	0.01
1: true	0.4	0.6

$P(X_2/X_1)$: Conditional probability of X_2 assuming that X_2 occurs

Figure 3. A basic element of a Bayesian network

causality of both nodes is defined as a Conditional Probability Table (CPT). The CPT defines the causal relationship between X_1 as a parent node and X_2 as a child node in the Bayesian network. The arrow between the nodes defines the causal relationship between the parent node and child node. The conditional damage probability of the child node $P(X_2/X_1)$ can be determined by the *CPT* in Fig.3.

For example, according to the marginalization [6], the probability $P(X_2=1)$ can be estimated as

$$P(X_2 = 1) = \frac{\sum_{X_1=0}^{1} P(X_2 = 1/X_1) \cdot P(X_1)}{\sum_{X_1=0}^{1} \sum_{X_2=0}^{1} P(X_2/X_1) \cdot P(X_1)} = 0.305 \tag{1}$$

When variable $X_1=1$ on the parent node is given, the probability $P(X_2=1/X_1=1)$ can be estimated as

$$P(X_2 = 1/X_1 = 1) = \frac{P(X_2 = 1/X_1 = 1) \cdot P(X_1 = 1)}{\sum_{X_2=0}^{1} P(X_2/X_1 = 1) \cdot P(X_1 = 1)} = 0.6 \tag{2}$$

On the contrary, when variable $X_2=1$ on the child node is given, the probability $P(X_1=1/X_2=1)$ can be estimated as

$$P(X_1 = 1/X_2 = 1) = \frac{P(X_2 = 1/X_1 = 1) \cdot P(X_1 = 1)}{\sum_{X_1=0}^{1} P(X_2 = 1/X_1) \cdot P(X_1)} = 0.984 \tag{3}$$

If the causal relationships among disaster events can be defined by a Bayesian network with two or more nodes, which includes the 2 nodes of Fig.3 as a minimum unit, the conditional probability of a target node can be improved with an increase in the observed information of the parent or child nodes.

3.3. Improvement of damage probability using Bayesian network

Fig.4 shows a proposed Bayesian network. The proposed Bayesian network consists of 4 nodes; (1) Earthquake Ground Motion ($EGM(a)$), (2)Electric Power Distribution Equipment damage ($EPDE$),(3) Damage Inspection(DI),and (4) Power Outage(PO). The CPT of the proposed Bayesian network is defined as follows.

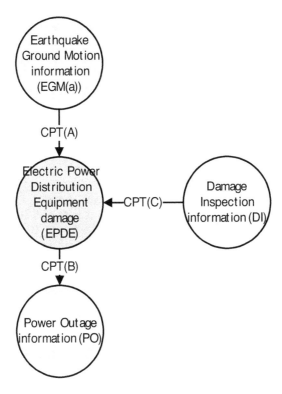

Figure 4. The proposed Bayesian network model

a. Causal relationship between earthquake ground motion and electric power distribution damage (CPT(A))

Table 1 shows the CPT which defines the casual relationship between earthquake ground motion and electric power distribution equipment damage shown as CPT(A) in Fig.4.

EGM(a)=1 indicates that the information of the maximum seismic ground motion intensity *a* for every target equipment is given. *EGM(a)* = *0* indicates that no earthquake ground motion information is given. *EPDE*=1 and *EPDE*=0 indicate the damage and no damage that occurs to equipment, respectively. $P_i(a)$ indicates the estimated damage rate of equipment *i* with the maximum seismic ground motion *a*. $P_i(a)$ is estimated by appendix[A]. This paper assumes that when an earthquake occurs, the maximum ground motion *a* is always given for every target equipment within several minutes. Thus, in Table 1, P_i (*EPDE*=1/*EGM(a)*=1) and P_i (*EPDE*=0/*EGM(a)*=1) are equivalent to $P_i(a)$ and 1-$P_i(a)$, respectively, and P_i(*EPDE*/*EGM(a)*=0) is neglected in Table 1.

CPT(A)	P_i(EPDE/EGM(a))	
	No Damage	Damage
EGM(a)	EPDE=0	EPDE=1
1:maximum seismic ground motion a occurs	1-P_i(a)	P_i(a)

Table 1. Conditional probability table between the earthquake intensity and the damage of electric power distribution equipment (CPT(A))

b.   Causal relationship between the electric power distribution damage and the power outage (CPT(B))

Table 2 shows the CPT which defines the causal relationship between the electric power distribution damage and the power outage shown as CPT(B) in Fig.4. According to CPT(B), when *EPDE* = 1 is given, which indicates that equipment *i* is damaged, the conditional no power outage probability of equipment *i*, P_i (*PO=0/EPDE=1*), and the conditional power outage probability of equipment *i*, P_i (*PO=1/EPDE=1*), are assumed to be 0 and 1, respectively.

On the other hand, when *EPDE*=0 is given, which indicates that equipment *i* is not damaged, the conditional power outage probability of equipment *i*, P_i (*PO=0/EPDE=0*), and no power outage probability of equipment *i*, P_i (*PO=1/EPDE= 0*), are, respectively, estimated as

$$P_i(PO = 0 / EPDE = 0) = \prod_{i=1}^{Ne-1} (1 - P_i(a)) \tag{4}$$

$$P_i(PO = 1 / EPDE = 0) = 1 - \prod_{i=1}^{Ne-1} (1 - P_i(a)) \tag{5}$$

where *Ne* indicates the total number of equipment connected to the same distribution line (the same feeder). Equation (4) and Equation (5) assumes that when equipment is damaged, a power outage occurs to all equipment connected to the distribution line of the damaged equipment.

CPT(B)	$P_i(PO/EPDE)$	
EPDE	No Power Outage $PO= 0$	Power Outage $PO=1$
0: No damage	Equation (4)	Equation (5)
1: Damage	0	1

Table 2. Conditional probability table between the damage to electric power distribution equipment and the power outage (CPT(B))

c. Causal relationship between damage inspection and electric power distribution equipment damage (CPT(C))

Table 3 shows the CPT which defines the causal relationship between the damage inspection and the electric power distribution equipment damage shown as CPT(C) in Fig.4. The damage inspection information indicates the inspection result of equipment with the same attribute as the target equipment i, which includes the number of damaged and non-damaged equipment on the condition that the target equipment i has no inspection. According to CPT(C), when $DI=0$ is given, which indicates that equipment i has no inspection information, the conditional damage probability of equipment i, P_i $(EPDE=1/DI=0)$, and the conditional no damage probability of equipment i, P_i $(EPDE=0/DI=0)$, are equivalent to $P_i(a)$ and $1- P_i(a)$, respectively.

On the other hand, when $DI=1$ is given, which indicates that equipment i has inspection information, the conditional damage probability of equipment i, P_i $(EPDE=1/DI=1)$, and the conditional no damage probability of equipment i, P_i $(EPDE= 0/DI=1)$, are evaluated as

$$P_i(EPDE = 1 / DI = 1) = \frac{n_i^0 + n_i^1 + 1}{M_i^0 + M_i^1 + 2} \tag{6}$$

$$P_i(EPDE = 0 / DI = 1) = 1 - \frac{n_i^0 + n_i^1 + 1}{M_i^0 + M_i^1 + 2} \tag{7}$$

where M_i^0 is the total number of inspected equipment with the same attribute as equipment i. n_i^0 is the total number of damaged equipment with the same attribute as equipment i. On the other hand, M_i^1 and n_i^1 are, respectively, evaluated as

$$M_i^1 = \frac{\mu_{p_i(a)}(1 - \mu_{p_i(a)})}{\sigma_{p_i(a)}^2} - 3 \tag{8}$$

$$n_i^1 = \mu_{p_i(a)} \left\{ \frac{\mu_{p_i(a)}(1 - \mu_{p_i(a)})}{\sigma_{p_i(a)}^2} - 1 \right\} - 1 \qquad (9)$$

where, $\mu_{p_i(a)}$ and $\sigma_{p_i(a)}$ are the average and standard deviation of the estimated damage probability of equipment with the same attribute as equipment I, $p_i(a)$, on the condition that the maximum ground motion intensity a is given, respectively. For this formulation, it is assumed that the estimated damage probabilities of equipment with the same attribute i follow the beta distribution based on the theory of Bayesian statistics [7].

Note that when the damage inspection information of equipment i is given, the conditional damage probability of equipment i, P_i ($EPDE/DI$=1), becomes 1 (damage) or 0 (no damage) as the definitive value instead of that by Table 3.

CPT(C)		$P_i(EPDE/DI)$
DI	No Damage EPDE=0	Damage EPDE=1
0: No inspection	1-$P_i(a)$	$P_i(a)$
1: Inspection	Equation (7)	Equation (6)

Table 3. Conditional probability table between the inspection information and the damage of electric power distribution equipment

d.   Combination of the joint occurrence probability

Based on Table1, Table2, and Table 3, Table 4 shows the combination of all the joint occurrence probabilities in Fig.4. Based on Table 4, all the conditional damage probabilities of the equipment can be evaluated. For example, when the earthquake ground motion information, $EGM(a)$=1 and power outage information, PO=1, are given, the conditional damage probability of equipment i, $P_i(EPDE$=1/$EGM(a)$=1,PO=1), is evaluated as

$$P_i(EPDE = 1 / EGM(a) = 1, PO = 1) = \frac{[1]+[5]}{[1]+[3]+[5]+[7]} \qquad (10)$$

On the other hand, when the earthquake ground motion information, $EGM(a)$=1 and the damage inspection information, DI=1, are given, the conditional damage probability of equipment i, $P_i(EPDE$=1/$EGM(a)$=1,DI=1), is evaluated as

$$P_i(EPDE = 1 / EGM(a) = 1, DI = 1) = \frac{[1]+[2]}{[1]+[2]+[3]+[4]} \qquad (11)$$

Combination No	EGM(a)	DI	EPDE	PO	Joint occurrence probability
[1]	1	1	1	1	Equation(6)
[2]	1	1	1	0	0
[3]	1	1	0	1	$(1-\text{Equation}(6))(1-\Pi(1-\text{Equation}(6)))$
[4]	1	1	0	0	$(1-\text{Equation}(7))\Pi(1-\text{Equation}(7))$
[5]	1	0	1	1	$P_i(a)$
[6]	1	0	1	0	0
[7]	1	0	0	1	$(1-P_i(a))(1-\Pi(1-P_i(a)))$
[8]	1	0	0	0	$(1-P_i(a))\Pi(1-P_i(a))$

Table 4. Combination of the joint probability (CPT(A)+CPT(B)+CPT(C))

The following positive analyses discuss the estimation accuracy for the situations of Equation (10) and Equation (11).

4. Positive analyses based on the 2007 Niigata-Ken Chuetsu Oki earthquake

4.1. Precondition of positive analyses

This chapter discusses the effectiveness of the proposed model based on an actual power outage and damage records of an electric power distribution system struck by the 2007 Niigataken-ken Chuetsu-Oki earthquake (hereafter, called the Chuestsu-Oki earthquake). The target electric power distribution system consists of 32,295 poles including 18,474 high voltage electric power distribution poles and 63 feeders, which indicates a high voltage distribution line.

Fig.5 shows the distribution of the seismic intensity scale of the Japan Meteorological Agency (JMA) due to the Chuetsu-Oki earthquake. The Chuetsu-Oki earthquake caused the 6 upper on the seismic intensity scale of the Japan Meteorological Agency (JMA 6+) as the maximum ground motion intensity to the struck area. In the analysis, it is assumed that the earthquake ground motion information related to Fig.5 has already been given as EGM(a) in Fig.4.

Fig.6 shows the power outage area caused by the Chestsu-Oki earthquake. Power outages also occurred to 15,074 high voltage poles, which is about 80 % of the total number of high voltage poles in the target system. The power outage information is given as the power outage information (PO) in Fig.4.

Fig.7 shows the observed points of the damaged poles due to the Chuetsu-Oki earthquake. The pole damages mainly consisted of two damage modes; i.e., breakage and inclination [8]. The damaged pole information is given as the damage inspection information (DI) in Fig.4.

In order to discuss the effectiveness of the proposed model, two situations are assumed in the following positive analysis;

1. The earthquake ground motion and power outage information are given (discussed in 4.2).

2. The earthquake ground motion and the damage inspection information are partially given
    (discussed in 4.3).

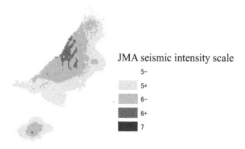

Figure 5. The distribution of the seismic intensity scale of the Japan Meteorological Agency (JMA) due to the Chuetsu-Oki earthquake (treated as EGM(a) in Fig.4)

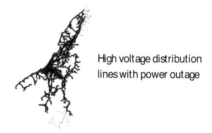

Figure 6. The observed power outage of high voltage distribution lines due to the Chuetsu-Oki earthquake (treated as PO in Fig.4)

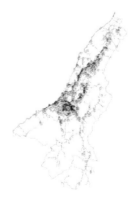

Figure 7. The observed points of the damaged poles due to the Chuetsu-Oki earthquake (treated as DI in Fig.4)

4.2. The effect on accuracy improvement of power outage information

Fig.8 shows a comparison between the observed and estimated number of damaged high voltage poles, which are normalized by the total number of observed damage poles. The caption [Observed] indicates the total number of observed poles damaged by the Chuetsu-Oki earthquake. The caption [without POI] indicates the total number of estimated damaged poles based on the causal relationship defined by Table 1, which is CPT (A) in Fig.4. The damage probability for pole i, which indicates the estimated damage number of pole i, is estimated as P_i ($EPDE= 1/EGM(a)=1$). On the other hand, the caption [with POI] indicates the total number of estimated damage poles based on the two causal relationships including CPT(A) and CPT(B) in Fig.4 evaluated by Equation (10).

Fig.8 indicates that the normalized damaged number of [with POI], 1.06, is closer to that of the [Observed], 1.00, than that of [without POI], 1.25. This result suggests that the proposed model using the power outage information can effectively improve the damage estimation accuracy of the electric power distribution poles.

Fig.9, on the other hand, shows a comparison of the number of damaged poles for every third mesh (1km×1km) among [Observed], [with POI], and [without POI]. R indicates the correlation coefficients between [Observed] and [with POI], and between [Observed] and [without POI]. Fig.9 indicates that the correlation coefficient R between [Observed] and [with POI] is slightly higher than that between [Observed] and [without POI].

In order to discuss the improved effect of the power outage information on the damage estimation accuracy, Fig.10 shows the relationship among the damage probability of a pole without power outage information, the damage probability of a pole with power outage information, and the total number of poles connected to the same feeder. The horizontal axis indicates the damage probability of a pole without power outage information (PO), which is estimated as P_i ($EPDE= 1/EGM$ (a)$=1$) based on Table 1. The vertical axis indicates that with POI, which is estimated as P_i ($EPDE=1/POI=1$) based on Equation (10). Ne, the total number of poles on the same feeder, imitates the actually installed feeder conditions of the target electric power distribution system.

Fig.10 illustrates that when the damage probability of a pole without POI is 0.001, POI improves the damage probability to 0.5, 0.1, 0.02 and 0.0015 on the condition that $Ne=2$, $Ne=10$, $Ne= 50$, $Ne=100$, and $Ne=1000$, respectively. This result suggests that the power outage information becomes more effective along with a decrease in the number of poles connected to the same feeder. In this paper, though it is assumed that a power company can identify a power outage range for every feeder, some power companies can identify the power outage within a subdivided range using a switch. In such a situation, the power outage information is more useful to improve the damage estimation accuracy.

Fig.10 also shows that the improvement effect based on the power outage information depends on the earthquake ground motion intensity level. For example, when the earthquake ground motion intensity under a target pole becomes about 6- to 6+ on the JMA seismic intensity scale, it is usually estimated that the damage probability without POI, which is evaluated by $P(a)$, P_i ($EPDE=1/EGM$ (a)$=1$), becomes 0.001 to 0.01. When the damage probability without POI is from

0.001 to 0.01, there are significant differences in the damage probability with POI. On the other hand, when $P(a)$ exceeds 0.03,whose earthquake ground motion intensity becomes over 6+, there is a limited effect of power outage information on improving the damage estimation accuracy.

This result suggests that the power outage information is usually effective for improving the estimation accuracy. However, when one feeder, which is a unit to identify the power outage range, has over 50 high voltage poles, and over 6+ of the earthquake ground motion level strike target feeder, there is a possibility that the effect of the power outage information only slightly improves the damage estimation accuracy of the target equipment.

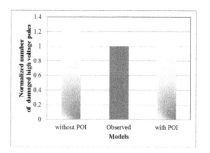

POI: Power Outage Information

Figure 8. Comparison of total number of damaged high voltage poles

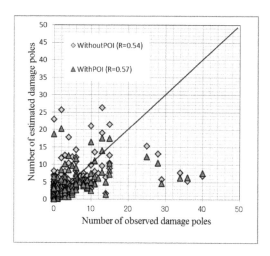

Figure 9. Comparison of the number of damaged poles for every third mesh (1km×1km) among [Observed], [with POI], and [without POI].

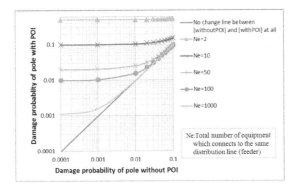

Figure 10. Effect of power outage information on the conditional damage probability of electric power distribution equipment

4.3. The effect on accuracy improvement by the damage inspection information

This section discusses the effect of the damage inspection information to improve the damage probability of the electric power distribution poles. In order to understand the effect of the inspection information on the improvement of the damage estimation accuracy, the damage probability of poles, P_i $(EPDE=1/EGM$ $(a)=1)$ is updated based on the different inspection rates. The inspection rate is evaluated as the number of inspected poles divided by the total number of poles (32,295 poles). In this simulation, it is assumed that actual damaged poles by the Chuetsu-Oki earthquake shown in Fig.7 are inspected based on an inspection priority. The inspection priority is determined as follows:

1. The target area is divided into the third mesh (1km×1km)

2. The third mesh (1km×1km) is also divided into the 16 fifth meshes (250m×250m).

3. The conditional damage probabilities, $P_i(EPDE=1/EGM(a)=1)$, for all poles are estimated based on table 1.

4. The averages of the estimated damage probabilities for every third mesh and for every fifth mesh are evaluated.

5. The differences in the average of all the estimated damage probabilities between the third meshes and the 16 fifth meshes of the same third mesh are evaluated as an inspection priority index.

6. The 16 fifth meshes are put in order based on the inspection priority index value. The smaller the priority index, the higher the priority level.

Fig.11 shows the allocated inspection districts of fifth meshes with a 29% inspection rate. The red square point shows the inspection point which is determined by the inspection priority index. The background chart with allocated third meshes is the estimated number of pole damages for every third mesh based on the conditional damage probabilities of all poles, $P_i(EPDE=1/EGM(a)=1)$.

Fig.12 compares the number of estimated damaged poles evaluated by Equation (11) to that of the observed damage poles for every third mesh. Fig.13 also shows a comparison of the number of estimated damaged poles to that of the observed damage poles for every third mesh. The estimated values consist of two factors; i.e., the number of estimated damaged poles with 29% inspection information and that without any inspection information. The correlation coefficient R between the number of estimated damaged poles with 29 % inspection informa- tion and that of the observed one is 0.89 while R between that without inspection information and the observed one is 0.59. This result suggests that the estimation accuracy is highly improved by the 29% inspection information.

Fig.14 shows a sensitivity analysis between the inspection rate and the correlation coefficient. Fig.14 indicates that when the inspection rate exceeds 0.35, the correlation coefficient R becomes greater than 0.9. This result suggests that when a target correlation coefficient is assumed, the districts to be inspected for obtaining target damage estimation accuracy could be automatically determined in a target electric power distribution system. The proposed model enables us to rationalize the inspection during the initial and emergency restoration periods.

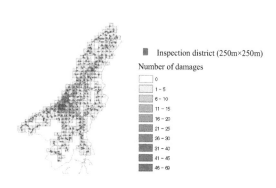

Figure 11. Effective inspection points for the damage estimation of a whole target area

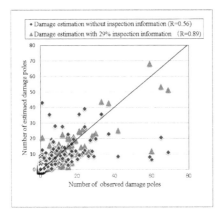

Figure 12. Comparison of estimation accuracies between the damage estimations with 29 % inspection information and without inspection information

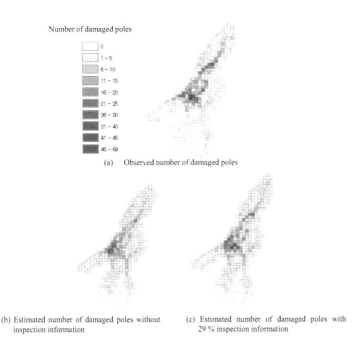

(b) Estimated number of damaged poles without
    inspection information

(c) Estimated number of damaged poles with
    29 % inspection information

Figure 13. Comparison of the number of estimated damaged poles to that of observed damaged poles for every third mesh.

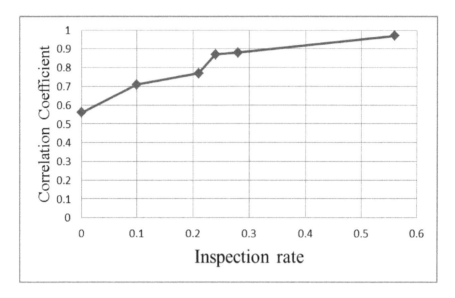

Figure 14. Relationship between the inspection rate and the correlation coefficient associated with the observed and estimation of damaged poles

5. Conclusion

This paper proposed a model to integrate multiple disaster information. The proposed model enables us to improve the damage estimation accuracy of electric power distribution equipment during the initial and emergency restoration periods after an earthquake. The research results are summarized as follows.

1. Information required for the emergency restoration work

The restoration process for an electric power distribution system after an earthquake is divided into three periods; i.e., initial, emergency, and permanent restoration periods. The necessary information and the information that was able to be collected within the three restoration periods were elucidated. As a result, it was clarified that the damage estimation technologies are very useful for actual restoration work under limited disaster information circumstances while the application of a seismic damage estimation system for electric power distribution equipment (RAMPEr) to the actual restoration work after the 2011 earthquake off the Pacific coast of the Tōhoku is described.

2. Formulation of a sequential updated model for electric power distribution system

A basic model to integrate the sequentially updated disaster information was proposed based on a Bayesian network. The proposed model can effectively integrate multidimensional disaster information, including earthquake ground motion, power outage and damage inspection information, to improve the estimation accuracy of seismic damaged electric power distribution poles.

3. Positive analysis based on the 2007 Niigata-Ken Chuetsu-Oki earthquake

The proposed model was applied to an actual electric power distribution system struck by the 2007 Niigataken Chuetsu-oki Earthquake, and the effect of the power outage and damage inspection information for improvement of the damage estimation accuracy was verified. As for the power outage information, it was clarified that under the installed conditions of the actual electric power distribution system, the damage estimation accuracy with power outage information was higher than that without power outage information. It was also realized that in order to effectively utilize the power outage information by the proposed model, the size of one feeder, which was related to a unit to identify the power outage range, and the earthquake ground motion level, which determined the damage probability level, were important parameters.

On the other hand, as for the inspection information, in order to effectively select the damage inspection point, the inspection priority of the actual electric power distribution poles was proposed. Based on the proposed inspection priority, the relationship between the inspection rate and the damage estimation accuracy was analyzed. As a result, it was also clarified that under the installed conditions of the actual electric power distribution system, the estimation accuracy is highly improved only by the 29% inspection information and when the inspection rate exceeds 0.35, the correlation coefficient R between the number of observed damaged poles and that of the estimated one becomes greater than 0.9.

In Japan, the occurrence of the Nankai Trough earthquake is feared. As mentioned in Chapter 2, the damage estimation system, RAMPEr, in which the proposed model has already been installed, is operating in some areas that could be highly affected by the Nankai Trough earthquake. In such a high seismic area, it is expected that RAMPEr will become a useful tool to support the restoration work after the earthquake. As future subjects, in order to improve the damage estimation accuracy of the proposed model, some remote sensing images will be integrated into the proposed model and the damage records due to the 2011 earthquake off the Pacific coast of Tohoku will be analyzed.

Apendix [A]

Equipment damage estimation model [4]

In this paper, based on reference [4], the seismic damage probability with the maximum earthquake ground motion a ($P_i(a)$) is evaluated as

$$P_i(a) = L_m \cdot C_i \cdot B_j \cdot T_k \cdot S_l \cdot f_c(z(a)) \tag{12}$$

where $f_c(z(a))$ is the seismic damage ratio of equipment i with seismic countermeasure C and seismic performance $z(a)$ assuming that the maximum ground motion a affects the target equipment. The seismic performance $z(a)$ is the seismic safety margin evaluated by the bending moment of the ground surface of the electric power distribution poles caused by the maximum ground motion a. S_l is the modification coefficient evaluated by the line connected type $l.T_k$ is the modification coefficient for the land use condition $k.B_j$ is the modification coefficient for the microtopography division j. C_i is the modification coefficient for the seismic countermeasure of equipment i. L_m is the modification coefficient for local region m. m is an electric power supply area covered by a business branch office.

Seismic performance $z(a)$ relative to the earthquake intensity a is evaluated as

$$z(a) = k1 \cdot k2 \cdot z_0(a) \tag{13}$$

where $z_0(a)$ is the safety margin relative to the maximum surface ground acceleration a (m/s^2), which defined as the ratio to the static earthquake force of the design collapse load (N). The static force is converted into the top concentration load of a distribution pole from the maximum surface ground acceleration.

According to the Japan Electric Technical Standards and Codes Committee (2007), $Z_0(a)$ is evaluated as

$$Z_0(e) = \begin{cases} \dfrac{K \cdot D_0 \cdot t^4}{120 P(e) \cdot (H + t_0)^2} & \text{(without pole anchor)} \\[2ex] \dfrac{0.3K(D_0 \cdot Q \cdot t^4 + AJ)}{P(e) \cdot (H + t_0{}')^2} & \text{(with pole anchor)} \end{cases} \tag{14}$$

K is a soil coefficient defined by the Japan Electric Technical Standards and Codes Committee (2007). K is divided into four types. The standard soil [A] is defined as 3.9×10^7 (N/m^4), which includes hardened soil, sand, gravel, and soil with small stones. The standard soil [B] is defined as 2.9×10^7 (N/m^4), which includes softer soil than [A]. Poor soil [C] is defined as 2.0×10^7 (N/m^4), which includes a kind of silt without soil. Poor soil [D] is defined as 0.8×10^7 (N/m^4), which includes moist clay and humid soil.

D_0 is the diameter on the ground surface of the distribution pole (m). t is the penetration depth of the distribution pole into the ground (m). H is the concentration load height from the ground surface (m).

$P(a)$ is the concentration load (kN) converted from the maximum surface ground acceleration a (m/s^2). $P(a)$ is evaluated as

$$P(a) = \frac{a}{H - 0.25} \times W_1 l_1 \tag{15}$$

where H is the height of the distribution pole from the ground surface (m), W_1 is mass of the upper ground part of the distribution pole (kg), l_1 is the height of the gravity center of the upper part of the distribution pole (m).

t_0 is the depth of the gyration center of the distribution pole from the ground surface, which is evaluated as

$$t_0 = \frac{2}{3}t \quad \text{(without pole anchor)} \tag{16}$$

$$t_0' = \frac{2}{3}\left(\frac{t^2 + 3nt_c^2}{t + 2nt_c}\right) \quad \text{(with pole anchor)} \tag{17}$$

$$n = \frac{A}{A_1} \tag{18}$$

where A is the area of the pole anchor. A_1 is the area of the base part of the distribution pole. A and A_1 are evaluated as

$$A = (L - D_0)d \tag{19}$$

$$A_1 = D_0 t \tag{20}$$

where L is the length of the pole anchor (m), d is the width of the pole anchor (m).

Q and J are evaluated as

$$Q = \frac{1}{12}(6m^2 - 8m + 3), \ m = \frac{t_0'}{t} \tag{21}$$

$$J = (t_0' - t_c)^2 t_c \tag{22}$$

where t_c is the depth of the center of the pole anchor from the ground surface (m).

$k1$ is a modification coefficient considering the effects of the overhead wire including strung and joint use wires, and overhead equipment such as an overhead transformer.

Based on a preliminary analysis, $k1$ is evaluated as

$$k1 = \frac{1}{a1 \times \sum_{i=1}^{4} w_i L_i + b1} \tag{23}$$

where $a1$ and $b1$ are recurrent coefficients evaluated by a preliminary analysis, and these values are assumed to be $a1$ is 0.000428, $b1$ is 1.0. Note that i=1: high voltage wire; i=2: low voltage wire; i=3: overhead transformer; i=4: joint use wire; w_i is the mass of I; L_i is the height of the distribution pole from the ground surface.

$k2$ is a modification coefficient considering adjacent distribution poles. $k2$ is assumed to be

$k2$=1.0 (in the case where adjacent distribution poles have no overhead equipment), $k2$=0.9 (in the case where the number of adjacent distribution poles with overhead equipment is less than three), $k2$ =0.85 (in the case where the number of adjacent distribution poles with overhead equipment is more than four).

In Equation (A.1), the damage ratio is evaluated as

$$f_c(z(a)) = \frac{DP_c}{TP_c} \cdot \frac{\ln_c(z(a) / \lambda_d, \xi_d)}{\ln_c(z(a) / \lambda_{all}, \xi_{all})} \tag{24}$$

$$\ln_c(z(a) / \lambda_x, \xi_x) = \frac{1}{\sqrt{2\pi} \cdot \xi_x \cdot z(a)} \cdot \exp\left[-\frac{1}{2} \cdot (\frac{\ln(z(a))) - \lambda_x}{\xi_x})^2 \right] \tag{25}$$

$$x = \left\{ \begin{array}{l} d \text{ (damaged equipment)} \\ all \text{ (all equipment)} \end{array} \right\}$$

where, $f_c(z(a))$ is the seismic damage ratio function of equipment with seismic countermeasure c and equipment performance $z(a)$ assuming that the maximum ground surface acceleration a. DP_c is the total number of actual seismic damaged equipment with seismic countermeasure c, TP_c is the total number of equipment with seismic countermeasure c. When x is d, $\ln_c(z(a)) / \lambda_{x=d}, \zeta_{x=d})$ is the log normal probability density function of the performance value $z(a)$ associated with damaged equipment with seismic countermeasure c due to a target earthquake. When x is d, $\lambda_{x=d}$ and $\zeta_{x=d}$ are the mean and standard deviation of $\ln(z(a))$ associated with the damaged equipment, respectively. When x is all, $\ln_l(z(a)); \lambda_{x=all}, \zeta_{x=all})$ is the log normal probability density function of the performance value z(a) of all equipment with seismic countermeasure c. $\lambda_{x=all}$ and $\zeta_{x=all}$ are the mean and standard deviation of $\ln(z(a))$ for all equipment with seismic countermeasure c, respectively.

Acknowledgements

The views and actual damage records expressed herein are based on research supported by several electric power companies including the Tohoku Electric Power Co., Inc.

Author details

Yoshiharu Shumuta

Civil Engineering Laboratory, Central Research Institute of Electric Power Industry, Chiba, Japan

References

[1] Japan Meteorological Agency: What is an Earthquake Early Warning ?, http://www.jma.go.jp/jma/en/Activities/eew1.html/ (accessed 21 November 2012).

[2] Sentinel Asia: https://sentinel.tksc.jaxa.jp/sentinel2/topControl.action/ (accessed 21 November 2012).

[3] Shumuta,Y., Todou, T., Takahashi, K., and Ishikawa, T. Development of a Damage Estimation Method for Electric Power Distribution Equipment, *The Journal of the Institute of Electrical Engineers in Japan*, Volume 130-C Number 7, 2010;1253-1261 (in Japanese).

[4] Shumuta,Y. Masukawa,K. Todou,T. and Ishikawa T. Proceedings of the 11th International Conference on Applications of Statistics and Probability in Civil Engineering(CASP11),2011;MS_218,2026-2033.

[5] Strong-motion Seismograph Networks (K-NET, KiK-net), National Research Institute for Earth Science and Disaster Prevention, http://www.k-net.bosai.go.jp/ (accessed 21 November 2012).

[6] Koski, T., and Noble, J. Bayesian Networks, Wiley series in probability and statistics: Wiley, 2009.

[7] Gelman,A.,Caelin, J.B.,Stern, H.S.,and Rubin, D.,B. Bayesian Data Analysis: Chapman & Hall/CRC,2004.

[8] The Niigata-Ken Chuetsu Earthquake Investigation Committee, Report of the 2004 Niigata-Ken Chuetsu-Oki earthquake damage investigation report, Japan Society of Civil Engineering, 2004 (in Japanese).

[9] Tohoku Electric Power Co. Inc. "Power outage information associated with the 2012 earthquake off the Pacific coast of Tōhoku", http://www.tohoku-epco.co.jp/information/1182212_821.html (accessed November 21, 2012).

[10] Miyako,T., and Todou, T,. Development of seismic damage estimation system for electric power equipment, *Electrical Review*, 2011; No.2011. 10, 76-78 (in Japanese).

Chapter 9

Daily Variation in Earthquake Detection Capability: A Quantitative Evaluation

Takaki Iwata

Additional information is available at the end of the chapter

1. Introduction

Evaluating the capability for detection of an earthquake catalogue is the first step in statistical seismicity analysis. In recent years, many studies have focused on ways to assess the completeness magnitude (M_C), the lowest magnitude level at which all earthquakes are recorded and there are no missing earthquakes, in a global earthquake catalogue [1, 2], regional catalogues [3–6], and small-scale observations such as underground mines [7]. We cannot make full use of the available information in an earthquake catalogue if we overestimate M_c, and underestimation of M_c yields incorrect or biased results.

[8] is a representative example showing that an accurate choice of M_C is vital in a seismicity analysis. This study examined the global earthquake catalogue provided by the National Oceanic and Atmospheric Administration (NOAA), United States Department of Commerce, and reported that the number of earthquakes during nighttime was significantly higher than that during daytime. Some studies [9, 10] stated that the daily variation in seismic activity [8] had found was just a bias; [8] did not take into account the daily change in detection capability, and earthquakes with small magnitudes — for which records were incomplete — were used in the analysis.

Empirically, it is well known that the level of cultural noise has daytime/nighttime variations and that this causes a daily variation in the detection capability. Nonetheless, in many studies, the estimation of M_c or the evaluation of the detection capability is carried out for a catalogue ranging over several days, weeks, months or years without taking short-time variation into consideration; M_c is underestimated in daytime and overestimated in nighttime, compared with the true value of M_c.

For an appropriate and precise evaluation of the detection capability, it is vital to quantify this temporal change. In this chapter, we propose a statistical method for a quantitative evaluation of the daily variation in the detection capability and also present an example of the method in real earthquake sequences. A short summary of this chapter has already been presented in a non-peer reviewed article [11], and this chapter focuses on showing the details of the approach and results that appeared in that article.

2. Data

The earthquake data used in this study were retrieved from the Japan Meteorological Agency (JMA) catalogue from January 2006 to December of 2010, with focal depths shallower than 30 km. The magnitude scale used in the JMA catalogue has been termed the "JMA magnitude", which is determined on the basis of velocity and displacement seismograms [12, 13].

The detection capability of earthquakes has regional variations, and in particular there is clear difference between inland and offshore regions. Because the main interest of this study is the temporal variation in the detection capability, to mitigate the influence of the regional differences, this study used the events occurring within the "Mainland" area defined by [14], which covers the inland and coastal regions of Japan (Figure 1). The total number of the events is 331,537.

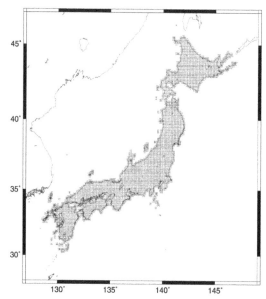

Figure 1. "Mainland" defined by [14], which covers the inland and coastal regions of Japan (after [14]).

Because the main object of the present study is to investigate the daily variation in earthquake detection capability, the data were divided into periods of one day, and the one-day sequences were stacked. The technical limitations of computer memory space and computing time make it difficult to handle hundreds of thousands events simultaneously in a Bayesian analysis, as described in the next section. Thus, the stacked sequences were constructed for each of the five years; five datasets were analyzed to examine the daily change in the detection capability.

3. Statistical method to evaluate the daily variation in detection capability of earthquakes

3.1. Evaluation of the detection capability by using a statistical model to represent an observed magnitude-frequency distribution

[15] proposed a probability density function $f(M)$ for an observed magnitude-frequency distribution of earthquakes over all magnitude range. The probability density function is assumed to be the product of two probability distributions.

The first comes from the Gutenberg-Richter (GR) law [16], which is equivalent to an exponential distribution:

$$w(M|\beta) = \beta \exp(-\beta M), \tag{1}$$

where the parameter β is related to the b-value of the GR-law and their relationship is described by $\beta = b \ln 10$.

The second is a detection rate function $q(M)$ showing the proportion of detected earthquakes to all earthquakes at magnitude M. Following the proposal by [17], the cumulative distribution function of a normal distribution is used as the detection probability function:

$$q(M|\mu,\sigma) = \int_{-\infty}^{M} \frac{1}{\sqrt{2\pi}\sigma} \exp\left[-\frac{(x-\mu)^2}{2\sigma^2}\right] dx \tag{2}$$

We normalize the product of the two functions, and then we obtain the target probability density function as follows (Figure 2):

$$
\begin{aligned}
f(M|\mu,\beta,\sigma) &= \frac{w(M|\beta)q(M|\mu,\sigma)}{\int_{-\infty}^{\infty} w(M|\beta)q(M|\mu,\sigma)dM} \\
&= \exp(-\beta M)q(M|\mu,\sigma) \cdot \beta \exp\left[\beta\mu - \frac{\beta^2\sigma^2}{2}\right].
\end{aligned} \tag{3}
$$

Of the three parameters (μ, β, and σ) in this statistical model, the parameter μ has the closest connection with the detection capability; μ indicates the magnitude at which the detection probability is 50 %. This means that the detection capability is better as the value of μ is smaller.

The values of the three parameters are estimated by the maximum likelihood method. The log-likelihood function ($\ln L$) is given by

$$\ln L(\beta,\mu,\sigma) = \sum_{i}^{N} \ln f(M_i|\beta,\mu,\sigma), \tag{4}$$

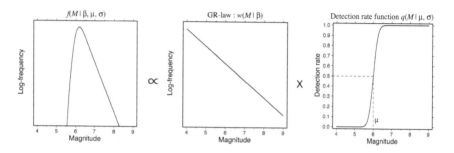

Figure 2. Schematic diagram showing the construction of the probability density function for an observed magnitude-frequency distribution of earthquakes over all magnitude range, as proposed by [15].

where N and M_i denote the number of the analyzed earthquakes and the magnitude of the i-th earthquake, respectively. The set of the values of the three parameters maximizing the log-likelihood function is the best estimate.

To observe the performance of this statistical model, we applied it to the magnitude-frequency distribution of the earthquakes taken from the JMA catalogue described in the previous section. As observed in Figure 3, the estimated curve of the statistical model fits well with the data, suggesting that this model is appropriate to describe the magnitude-frequency distribution on the JMA magnitude scale.

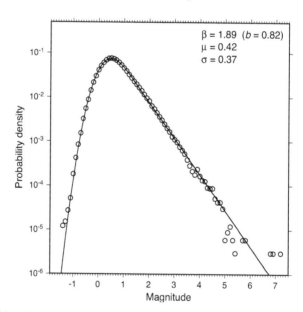

Figure 3. Observed (open circles) and estimated (solid line) probability density functions of the magnitudes of all the shallow (depth $\leq$ 30 km) earthquakes within the "Mainland" area [14], from January 2006 to December 2010. The best estimates of β (or b-value), μ, and σ are shown in the top right corner.

As a preliminary analysis to examine the daily variation, this model was applied to the datasets of nighttime (0.0–0.2 days) and daytime (0.4–0.6 days) earthquakes; we found a clear shift of the detection capability between the two datasets (Figure 4).

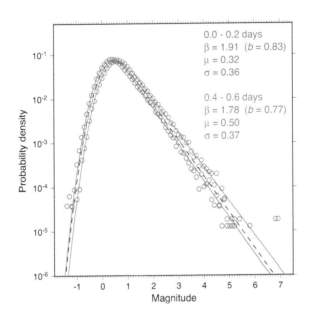

Figure 4. Observed (open circles) and estimated (solid line) probability density functions of the magnitude of nighttime (0.0–0.2 days, red) and daytime (0.4–0.6 days, blue) earthquakes. The dotted black line indicates the estimated probability density function for all the shallow earthquakes used in this study, as shown in Figure 3.

3.2. Evaluation of the daily variation of the detection capability in a Bayesian framework

To quantify the daily variation of the detection capability, we estimated the temporal change in the three parameters $\mu, \beta,$ and σ that appeared in the model. The procedure of the estimation is similar to that used in [2, 18].

To represent the temporal variations in the parameters, we introduced a piecewise linear function or linear spline [19]. The nodal point of the spline were taken at each of the occurrence times of the earthquakes in a stacked sequence. For the sake of brevity, we defined $\theta_i^{(1)}, \theta_i^{(2)},$ and $\theta_i^{(3)}$ as the values of $\beta, \mu,$ and σ, respectively, at the occurrence time t_i of the i-th earthquake. Hence, the temporal variations of $\mu(t), \beta(t),$ and $\sigma(t)$ were represented by

$$\phi_1(t) = \beta(t) = \frac{\theta_{i+1}^{(1)} - \theta_i^{(1)}}{t_{i+1} - t_i}(t - t_i) + \theta_i^{(1)},$$

$$\phi_2(t) = \mu(t) = \frac{\theta_{i+1}^{(2)} - \theta_i^{(2)}}{t_{i+1} - t_i}(t - t_i) + \theta_i^{(2)},$$

$$\phi_3(t) = \sigma(t) = \frac{\theta_{i+1}^{(3)} - \theta_i^{(3)}}{t_{i+1} - t_i}(t - t_i) + \theta_i^{(3)} \quad \text{for } t_i \le t < t_{i+1}. \tag{5}$$

The goal of the estimation is the optimization of the following parameters:

$$\boldsymbol{\theta} = (\boldsymbol{\theta}^{(1)}; \boldsymbol{\theta}^{(2)}; \boldsymbol{\theta}^{(3)})$$
$$= (\theta_1^{(1)}, \dots, \theta_N^{(1)}; \theta_1^{(2)}, \dots, \theta_N^{(2)}; \theta_1^{(3)}, \dots, \theta_N^{(3)}). \tag{6}$$

The simultaneous optimization of such a large number ($= 3N$) of parameters is, however, an unstable process; we incorporated a smoothness constraint or roughness penalty on $\phi_i(t)$ ($i = 1, 2, 3$) to enhance the stability of the optimization.

Following the equation (4), the log-likelihood function in this case is given by

$$\ln L(\boldsymbol{\theta}) = \sum_i^N \ln f(M_i | \theta_i^{(1)}, \theta_i^{(2)}, \theta_i^{(3)}), \tag{7}$$

and the smoothness constraint is quantified by the following equation:

$$\Phi(\boldsymbol{\theta}|v_1, v_2, v_3) = \sum_{j=1}^{3} v_j \int_{T_s}^{T_e} \left[\frac{\partial}{\partial t} \phi_j(t) \right]^2 dt, \tag{8}$$

where v_j is the parameter controlling the trade-off between the goodness-of-fit of the statistical model to the data and the smoothness constraint, and $[T_s, T_e]$ denotes the domain of the analyzed time period.

In the analysis of an ordinary earthquake sequence such as [2, 18], the start and end points (T_s and T_e) are different but in the present case they are the same, midnight because the stacked one-day data are analyzed. Thus, the values of $\beta, \mu,$ and σ at T_s and T_e should connect smoothly. Considering this property and the equation (5) we rewrite the equation (8) as follows:

$$\Phi(\boldsymbol{\theta}|v_1, v_2, v_3) = \sum_{j=1}^{3} v_j \left[\sum_{i=1}^{N-1} \frac{(\theta_{i+1}^{(j)} - \mu_i^{(j)})^2}{t_{i+1} - t_i} + \frac{(\theta_1^{(j)} - \theta_N^{(j)})^2}{(t_1 - T_s) + (T_e - t_N)} \right] \tag{9}$$

Then, we introduced a penalized log-likelihood function [20, 21]

$$Q(\boldsymbol{\theta}|v_1, v_2, v_3) = \ln L(\boldsymbol{\theta}) - \Phi(\boldsymbol{\theta}|v_1, v_2, v_3), \tag{10}$$

and the maximization of $Q(\boldsymbol{\theta}|v_1, v_2, v_3)$ provides the best estimate of $\boldsymbol{\theta}$, and it depends on the values of $v_j (j = 1, 2, 3)$.

A Bayesian framework with the type II maximum likelihood approach [22] or the maximization of marginal likelihood [23] enables us to determine the values of v_j's objectively. We supposed that the prior distribution of $\boldsymbol{\theta}$ corresponding to the smoothness constraint is proportional to $\exp[-\Phi(\boldsymbol{\theta}|v_1, v_2, v_3)]$; the equation (9) and the consideration of the normalizing constant give us the prior distribution $\pi(\boldsymbol{\theta}|v_1, v_2, v_3)$ as follows:

$$
\begin{aligned}
&\pi(\boldsymbol{\theta}|v_1, v_2, v_3) \\
&= \prod_{j=1}^{3} \prod_{i=1}^{N-1} \sqrt{\frac{v_j}{\pi(t_{i+1} - t_i)}} \exp\left[-\frac{v(\phi_{i+1}^{(j)} - \phi_i^{(j)})^2}{t_{i+1} - t_i}\right] \\
&\quad \cdot \sqrt{\frac{v_j}{\pi(t_1 - T_s + T_e - t_N)}} \exp\left[-\frac{v(\phi_1^{(j)} - \phi_N^{(j)})^2}{t_1 - T_s + T_e - t_N}\right].
\end{aligned}
\tag{11}
$$

To find the marginal likelihood [24], we need to integrate the product of the prior distribution $\pi(\boldsymbol{\theta}|v_1, v_2, v_3)$ and the likelihood function $L(\boldsymbol{\theta})$ shown in the equation (7) over $\boldsymbol{\theta}$, but this integration is unachievable. This is because the integral of $\pi(\boldsymbol{\theta}|v_1, v_2, v_3)$ over $\boldsymbol{\theta}$ is infinite and is a so-called improper prior.

To deal with this computational problem we isolated $\theta_N^{(j)} (j = 1, 2, 3)$ from $\boldsymbol{\theta}$, because the integral of $\pi(\boldsymbol{\theta}|v_1, v_2, v_3)$ over

$$
\begin{aligned}
\boldsymbol{\theta}_{-N} &= (\boldsymbol{\theta}_{-N}^{(1)}; \boldsymbol{\theta}_{-N}^{(2)}; \boldsymbol{\theta}_{-N}^{(3)}) \\
&= (\theta_1^{(1)}, \dots, \theta_{N-1}^{(1)}; \theta_1^{(2)}, \dots, \theta_{N-1}^{(2)}; \theta_1^{(3)}, \dots, \theta_{N-1}^{(3)}).
\end{aligned}
\tag{12}
$$

is finite; we rewrote the original prior distribution $\pi(\boldsymbol{\theta}|v_1, v_2, v_3)$ as $\pi_{-N}(\boldsymbol{\theta}_N|v_1, v_2, v_3, \theta_N^{(1)}, \theta_N^{(2)}, \theta_N^{(3)})$ which is a proper prior with respect to $\boldsymbol{\theta}_{-N}$.

Then, to obtain the marginal likelihood $\mathcal{L}$, we integrated out the product of the prior distribution $\pi_{-N}(\boldsymbol{\theta}_{-N}|v_1, v_2, v_3, \theta_N^{(1)}, \theta_N^{(2)}, \theta_N^{(3)})$ and the likelihood function $L(\boldsymbol{\theta})$ over $\boldsymbol{\theta}_{-N}$:

$$\mathcal{L}(v_1, v_2, v_3, \theta_N^{(1)}, \theta_N^{(2)}, \theta_N^{(3)}) = \int_{\Theta} L(\boldsymbol{\theta})\pi_{-N}(\boldsymbol{\theta}_{-N}|v_1, v_2, v_3, \theta_N^{(1)}, \theta_N^{(2)}, \theta_N^{(3)})d\boldsymbol{\theta}_{-N}, \tag{13}$$

where Θ denotes the parameter space of $\boldsymbol{\theta}_{-N}$.

We intended to find the set of the values of $v_1, v_2, v_3, \theta_N^{(1)}, \theta_N^{(2)}$, and $\theta_N^{(3)}$ which maximizes the marginal likelihood, because such a set is the best estimate of the six parameters [22, 23].

In this hierarchical Bayesian scheme, the six parameters are often called as hyperparemeters which govern the prior distribution.

The maximization of the marginal likelihood $\mathcal{L}(v_1, v_2, v_3, \theta_N^{(1)}, \theta_N^{(2)}, \theta_N^{(3)})$ was achieved in the following way. In the first step, we intend to maximize the integrand $\ln L(\boldsymbol{\theta})\pi_{-N}(\boldsymbol{\theta}_{-N}|v_1, v_2, v_3, \theta_N^{(1)}, \theta_N^{(2)}, \theta_N^{(3)})$ in the equation (13) with respect to $\boldsymbol{\theta}_{-N}$, which is equivalent to the maximization of the penalized log-likelihood function $Q(\boldsymbol{\theta}|v_1, v_2, v_3)$ appeared in the equation (10).

The logarithm of the integrand in the equation (13) is approximated by a quadratic form at the initial value of $\boldsymbol{\theta}_{-N} = \boldsymbol{\theta}_{-N0}$ (and $\boldsymbol{\theta} = \boldsymbol{\theta}_0$):

$$
\begin{aligned}
&\ln L(\boldsymbol{\theta})\pi_{-N}(\boldsymbol{\theta}_{-N}|v_1, v_2, v_3, \theta_N^{(1)}, \theta_N^{(2)}, \theta_N^{(3)})\\
&\approx \ln L(\boldsymbol{\theta}_0)\pi_{-N}(\boldsymbol{\theta}_{-N0}|v_1, v_2, v_3, \theta_N^{(1)}, \theta_N^{(2)}, \theta_N^{(3)})\\
&+ g(\boldsymbol{\theta}_{-N0}|v_1, v_2, v_3, \theta_N^{(1)}, \theta_N^{(2)}, \theta_N^{(3)}) \cdot (\boldsymbol{\theta}_{-N} - \boldsymbol{\theta}_{-N0})\\
&- \frac{1}{2}(\boldsymbol{\theta}_{-N} - \boldsymbol{\theta}_{-N0})H(\boldsymbol{\theta}_{-N0}|v_1, v_2, v_3, \theta_N^{(1)}, \theta_N^{(2)}, \theta_N^{(3)})(\boldsymbol{\theta}_{-N} - \boldsymbol{\theta}_{-N0})^T,
\end{aligned} \tag{14}
$$

where $g(\boldsymbol{\theta}_{-N0}|v_1, v_2, v_3, \theta_N^{(1)}, \theta_N^{(2)}, \theta_N^{(3)})$ and $H(\boldsymbol{\theta}_{-N0}|v_1, v_2, v_3, \theta_N^{(1)}, \theta_N^{(2)}, \theta_N^{(3)})$ are the gradient vector and the negative of the Hessian matrix (second partial derivatives) of the integrand at $\boldsymbol{\theta}_{-N} = \boldsymbol{\theta}_{-N0}$, respectively. The symbol T denotes the transpose of a vector (or a matrix).

If the choice of the initial value $\boldsymbol{\theta}_{-N0}$ is appropriate and is not far from $\boldsymbol{\theta}_{-N} = \hat{\boldsymbol{\theta}}_{-N}$ which maximizes the integrand, the quadratic form is an upward convex and $H(\boldsymbol{\theta}_{-N0}|v_1, v_2, v_3, \theta_N^{(1)}, \theta_N^{(2)}, \theta_N^{(3)})$ is a positive definite. Hence, Cholesky decomposition [25] enables us to factorize the matrix $H(\boldsymbol{\theta}_{-N0}|v_1, v_2, v_3, \theta_N^{(1)}, \theta_N^{(2)}, \theta_N^{(3)})$ into the following form:

$$
H(\boldsymbol{\theta}_{-N0}|v_1, v_2, v_3, \theta_N^{(1)}, \theta_N^{(2)}, \theta_N^{(3)}) = AA^T, \tag{15}
$$

where A is a lower triangular matrix.

The quadratic form is maximized at $\boldsymbol{\theta}_{-N} = \boldsymbol{\theta}'_{-N0}$ which satisfies

$$
A^T(\boldsymbol{\theta}_{-N} - \boldsymbol{\theta}_{-N0}) = A^{-1}g(\boldsymbol{\theta}_{-N0}|v_1, v_2, v_3, \theta_N^{(1)}, \theta_N^{(2)}, \theta_N^{(3)}) \tag{16}
$$

through the Newton method. We replace $\boldsymbol{\theta}_{-N0}$ by $\boldsymbol{\theta}'_{-N0}$ and iterate the procedure until $\boldsymbol{\theta}_{-N}$ converges to find $\boldsymbol{\theta}_{-N} = \hat{\boldsymbol{\theta}}_{-N}$.

At $\boldsymbol{\theta}_{-N} = \hat{\boldsymbol{\theta}}_{-N}$ where the integrand $\ln L(\boldsymbol{\theta})\pi_{-N}(\boldsymbol{\theta}_{-N}|v_1, v_2, v_3, \theta_N^{(1)}, \theta_N^{(2)}, \theta_N^{(3)})$ is maximized, the second term of the equation (14) vanishes because $g(\hat{\boldsymbol{\theta}}_{-N}|v_1, v_2, v_3, \theta_N^{(1)}, \theta_N^{(2)}, \theta_N^{(3)})$ is a zero vector. Hence, around $\boldsymbol{\theta}_{-N} = \hat{\boldsymbol{\theta}}_{-N}$, the integrand is approximated by

$$\ln L(\boldsymbol{\theta}) \pi_{-N}(\boldsymbol{\theta}_{-N} | v_1, v_2, v_3, \theta_N^{(1)}, \theta_N^{(2)}, \theta_N^{(3)})$$
$$\approx \ln L(\hat{\boldsymbol{\theta}}_0) \pi_{-N}(\hat{\boldsymbol{\theta}}_{-N} | v_1, v_2, v_3, \theta_N^{(1)}, \theta_N^{(2)}, \theta_N^{(3)})$$
$$- \frac{1}{2}(\boldsymbol{\theta}_{-N} - \hat{\boldsymbol{\theta}}_{-N}) H(\hat{\boldsymbol{\theta}}_{-N} | v_1, v_2, v_3, \theta_N^{(1)}, \theta_N^{(2)}, \theta_N^{(3)})(\boldsymbol{\theta}_{-N} - \hat{\boldsymbol{\theta}}_{-N})^T \qquad (17)$$

from the equation (14).

Then, in the second step, the integral in the equation (13) is computed by the Laplace approximation [26]. Consequently, the log marginal likelihood $\ln \mathcal{L}(v_1, v_2, v_3, \theta_N^{(1)}, \theta_N^{(2)}, \theta_N^{(3)})$ is given by

$$\ln \mathcal{L}(v_1, v_2, v_3, \theta_N^{(1)}, \theta_N^{(2)}, \theta_N^{(3)})$$
$$\approx \ln L(\hat{\boldsymbol{\theta}}_{-N}) \pi(\hat{\boldsymbol{\theta}}_{-N} | v_1, v_2, v_3, \theta_N^{(1)}, \theta_N^{(2)}, \theta_N^{(3)})$$
$$- \frac{1}{2} \ln \det H(\hat{\boldsymbol{\theta}}_{-N} | v_1, v_2, v_3, \theta_N^{(1)}, \theta_N^{(2)}, \theta_N^{(3)}) + \frac{n}{2} \ln 2\pi, \qquad (18)$$

where n is the number of parameters included in $\hat{\boldsymbol{\theta}}_{-N}$ and $n = 3(N-1)$ in this case.

By repeating the two steps, we attempt to find the values of the six hyperparameters which maximize the value of $\ln \mathcal{L}(v_1, v_2, v_3, \theta_N^{(1)}, \theta_N^{(2)}, \theta_N^{(3)})$. Eventually, we can find the optima of the hyperparameters and $\boldsymbol{\theta}_{-N}$.

With a sufficiently large n, the quadratic and Laplace approximations are good [27] and $\ln L(\boldsymbol{\theta}) \pi_{-N}(\boldsymbol{\theta}_{-N} | v_1, v_2, v_3, \theta_N^{(1)}, \theta_N^{(2)}, \theta_N^{(3)})$ is approximately a multidimensional Gaussian distribution. In this case, the inverse of the negative of the Hessian matrix $H(\hat{\boldsymbol{\theta}}_{-N} | v_1, v_2, v_3, \theta_N^{(1)}, \theta_N^{(2)}, \theta_N^{(3)})^{-1}$ gives the estimation errors of the parameters [15]; the $[i + j(N-1)]$-th diagonal component of $H(\hat{\boldsymbol{\theta}}_{-N} | v_1, v_2, v_3, \theta_N^{(1)}, \theta_N^{(2)}, \theta_N^{(3)})^{-1}$ is the standard error of $\theta_i^{(j)}$.

3.3. Model comparison

So far, we assumed that each of the parameters β, μ, and σ has daily variations. For more appropriate statistical modelling, however, it is necessary to examine the significance of the temporal variations of the three parameters.

To apply the model where at least one of the three parameters β, μ, and σ is not assumed to have the daily variation to data, we fix v_j('s) at 0 where j takes the value(s) corresponding to the parameter(s) without the temporal variation. Then the procedure described in section 3.2 is conducted, but we exclude $\theta^{(j)}$ in the equation (6) and consider this exclusion in the procedure following the equation.

The possbile combination of the allowance or non-allowance of the temporal variation in β, μ, and σ yields eight cases in total. We introduce Akaike's Bayesian Information Criterion ABIC [23] in the comparison of the goodness-of-fit of the eight cases to data:

$$ABIC = -2(\text{maximum } \ln \mathcal{L}(v_1, v_2, v_3, \theta_N^{(1)}, \theta_N^{(2)}, \theta_N^{(3)})$$
$$+2(\text{number of non-fixed hyperparameters}), \tag{19}$$

where $\ln \mathcal{L}(v_1, v_2, v_3, \theta_N^{(1)}, \theta_N^{(2)}, \theta_N^{(3)})$ is the marginal likelihood given by the equation (18). The model with smaller ABIC value is considered to have a better fit to data.

4. Results of the analysis of the JMA catalogue

On the basis of the comparison of the ABIC values for the eight models, we found that not only μ but also β and/or σ have the significant daily changes in the four stacked sequences except in the sequence for 2009.

Figure 5 shows the estimated daily variation of β (or b-value) found in the sequence for 2007. In this sequence the value of β drastically decreases at around 0.4 days (i.e., approximately 9:30 a.m.). On 27 March 2007 and 16 July 2007, there were two major earthquakes, the Noto Hanto earthquake with $M_w = 6.9$ and the Chuuetsu-oki earthquake with $M_w = 6.6$. Their occurrence times are coincidentally close to 10 a.m. (9:41 a.m. and 10:13 a.m. for the Noto Hanto and Chuuetsu-oki earthquakes, respectively), and these two major earthquakes are followed by active aftershock sequences (see Figure 5b). Thus, it is suspicious that the drastic decrease of β is the daily variation of our interest in this study and the two aftershock sequences would cause the decrease of β. To confirm this point the aftershock sequences following the Noto Hanto and Chuuetsu-oki earthquakes were excluded, and then the same analysis was applied; the model where only the daily variation of μ is allowed was chosen as the best model.

Year	Excluded sequence and area
2006	Earthquake swarm in Izu Peninsula (139.00-139.45°E, 34.65-35.15°N)
2007	The Noto Hanto earthquake and its aftershock sequence (136.60-137.00°E, 37.00-37.50°N)
	The Chuuetsu-oki earthquake and its aftershock sequence (138.30-138.80°E, 37.30-37.65°N)
2008	The Iwate-Miyagi Nairiku earthquake and its aftershock sequence (140.50-141.10°E, 38.70-39.30°N)
2010	Earthquake sequence in Nakadoori, Fukushima Prefecture (139.95-140.10°E, 37.20-37.35°N)

Table 1. Active sequences which may affect the examination of the daily variation in β, μ, and σ. The corresponding areas are shown in parentheses.

In similar to this case, from the sequences for 2006, 2008, and 2010, the active sequences listed in Table 1 were excluded and we re-analyzed them after the exclusion. Consequently, the case with only the daily variation of μ was considered as the best case for these three

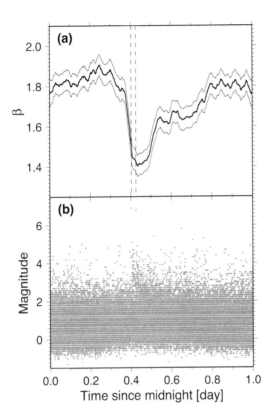

Figure 5. (a) Estimated daily variation of β for the stacked sequence for 2007 (bold line), and its two standard error bands (thin lines), as a function of the elapsed time since midnight. The two dotted lines indicate the occurrence times of the Noto Hanto and Chuuetsu-oki earthquakes. (b) Magnitude-time plot of the stacked sequence for 2007.

years, the same as that for 2007 (Table 2). As previously mentioned, in the original sequence for 2009, the case where only the daily variation of μ is allowed has been chosen as the best one. This is probably because no significant seismic activity occurred in that year.

Figure 6 depicts the daily variations of μ estimated for the sequences obtained in each of the five years with the exclusion of the significant activities listed in Table 1 (except 2009). There exist minor differences in these five profiles of μ, but they follow the almost the same pattern. First, during the time period between 0.0 and 0.2 days (i.e., between 0 a.m. and 5 a.m.) μ takes the smallest value and the detection capability is the best. Then the value of μ gradually increases as the time elapses, and it has a local peak around 0.4 days. Following the local peak μ shows a transient decrease corresponding to the recovery of the detection capability during lunchtime. At around 0.6 days (between 2 p.m. and 3 p.m.) it has a local peak again, and then it becomes smaller as the time gets closer to midnight. The differences of the maximum and minimum values of μ are 0.21–0.24.

Models			ΔABIC				
temporal variation			Year				
β	μ	σ	2006	2007	2008	2009	2010
			1160.9	1125.1	1092.6	1405.3	1448.9
✓			266.7	320.3	236.5	495.7	383.5
	✓		0.0	0.0	0.0	0.0	0.0
		✓	389.8	352.2	336.1	450.91	546.6
✓	✓		2.0	1.8	2.0	2.0	2.0
✓		✓	127.4	139.4	103.8	183.93	208.6
	✓	✓	2.0	2.0	2.0	1.1	1.1
✓	✓	✓	3.9	4.0	4.0	1.4	1.5

Table 2. Model comparison through ABIC values. The differences in ABIC values, ΔABIC, from cases where only the daily variation in μ is allowed are shown. The mark ✓ indicates the allowance of the daily variation in β, μ, and σ.

5. Discussions

To some extent, we can figure out the temporal pattern of the detection capability through the magnitude-time plots in Figure 6. In these plots there are some absences of small earthquakes with $M = -0.5$ or below during daytime. These plots also show a short recovery of the detection capability around noon. [28] has found a similar pattern of the detection capability in California through the hourly distribution of the number of reported earthquakes in the Advanced National Seismic Systems (ANSS) catalogue. Note that, however, both magnitude-time plots and hourly distributions provides us with only a qualitative pattern of the detection capability. For an understanding of the precise characteristics of an earthquake catalogue it is necessary to evaluate quantitatively the temporal change of the detection capability, and the statistical approach shown in this chapter is effective for such a purpose.

Of the three parameters β (or b-value), μ, and σ in the statistical model to describe an observed magnitude-frequency distribution, only μ shows the significant daily variation whereas β and σ do not show any variation. The temporal changes of b-value have been reported in some studies [15, 29, 30], and laboratory experiments [31–33], depth dependence [34–36], and faulting-style dependence of the b-value [37, 38] suggest a relation between stress state and b-value. We could consider air temperature and atmospheric pressure as physical factors which may cause the diurnal cycle of the stress changes in the Earth's crust. However, a constant value of β implies that the diurnal stress changes following the daily change of the temperature and/or pressure are insufficient to affect the magnitude-frequency distribution of earthquakes.

It is difficult to arrive at a plausible interpretation for a constant value of σ, because the factors controlling the value of σ still remain unknown. It has been suggested that σ is related to the spatial distribution of seismograph stations [2]. Following this suggestion the result of the unchanged σ is reasonable, because the spatial distribution of seismograph stations does not have daily variation.

As briefly described in section 4, the differences between the maximum and minimum values of μ are less than 0.25. In Figure 6, the values of μ in the case where we do not consider the daily variation of μ are indicated by the dotted lines; the difference between the maximum value of μ with the daily variation and that without the variation is approximately 0.15.

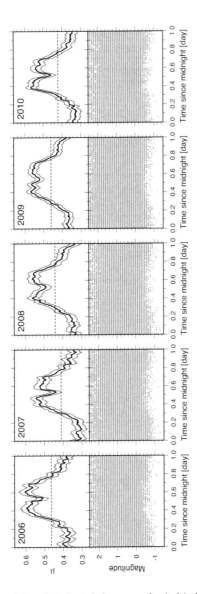

Figure 6. (Top) Estimated daily variations of μ in the stacked sequences of each of the five years (bold line) with the exclusion of the significant activities listed in Table 1, and their two standard error bands (thin lines), as a function of the elapsed time since midnight. The horizontal dotted lines shows the best estimate of μ in the case where we do not consider the temporal variation of μ. (Bottom) Magnitude-time plot of the stacked sequences for 2007. Note that only earthquakes with $M \leq 2.5$ are plotted to clarify the absences of small earthquakes during daytime but earthquakes with $M > 2.5$ are used in the analysis.

This suggests that, if we determine M_c without taking into consideration the daily variation of the detection capability, $M_c - 0.2$ would be appropriate as the completeness magnitude to avoid any unexpected biases in our seismicity analysis.

In the present study, we use the smoothness constraint, as shown in the equation (9), which reflects the cyclic property of data. Suppose that we use a different smoothness constraint

$$\Phi(\theta|v_1, v_2, v_3) = \sum_{j=1}^{3} v_j \left[\sum_{i=1}^{N-1} \frac{(\theta_{i+1}^{(j)} - \mu_i^{(j)})^2}{t_{i+1} - t_i} \right], \tag{20}$$

where we do not impose a smoothness constraint between the start and end points on the values of β, μ, and σ. If we analyze a sufficiently large dataset such as the five sequences used in this study, the introduction of the constraint without considering the cyclic property is not problematic. This is because a sufficiently large sample size reasonably reduce random fluctuations or estimation errors, and therefore the estimated values of the parameters at the start and end points connects smoothly without the smoothness constraint. In actual, the analyses with the constraints given by the equations (20) and (9) gives almost identical temporal profiles of μ for the five sequences.

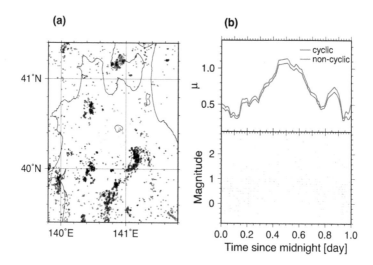

Figure 7. (a) Map showing the 300 earthquakes (green circles) used in the demonstrative analysis. (b) (top) Estimated daily variations of μ using the constraint following the equation (9) (red) and those following the equation (20). (bottom) Magnitude-time plot of the one-day stacked sequence of the 300 earthquakes.

For a small dataset, however, it is important to use the appropriate smoothness constraint. As a demonstration of the importance, a dataset containing 300 earthquakes distributed locally in a northern part of Japan (Figure 7a) was analyzed using the two constraints. The estimated daily variation of μ is shown in Figure 7b; we can observe systematic deviation between the two cases. The ABIC value with the smoothness constraint of the equation (9) is smaller than that corresponding to the equation (20), and the difference between the two ABIC values is 2.5, which is statistically significant. This demonstration implies that an analysis with an unreasonable constraint or prior distribution, which does not involve the essential characteristic of data and is statistically invalid, may leads us to an incorrect assessment of detection capability.

6. Concluding remarks

This chapter provided a statistical technique to make a quantitative evaluation of the daily variation of earthquake detection capability. As an example of its application to actual data, the datasets taken from the recent JMA catalogue were analyzed, and a guideline for the choice of completeness magnitude in the JMA catalogue was shown.

It should be noted that the results shown in this chapter were derived from an analysis where the earthquake sequence over an wide area was analyzed simultaneously. The daily variation of cultural noise, however, must have location dependency. Thus, in a manner similar to the analysis shown in Figure 7, we should examine regional earthquake sequences comprehensively to deal with an earthquake catalogue in a more sophisticated manner.

In this chapter we considered only the daily variation of detection capability, but human activities also have a weekly periodicity [28]. Additionally, we often observe seasonal (annual) variations in seismic noise level [39, 40], which are mainly caused by meteorological factors. Thus, the development of an appropriate method to handle such multiple periodic variations of earthquake detection capability is a necessary challenge to determine the completeness magnitude more accurately.

Acknowledgments

This study was partially supported by the Grants-in-Aid 23240039 for Scientific Research (A) by the Ministry of Education, Culture, Sports, Science and Technology, Japan, and by the ISM Cooperative Research Program (2011-ISM·CRP-2007). To obtain the results shown in this chapter, the author used the supercomputer system of the Institute of Statistical Mathematics, Japan. Figures were generated using the GMT software [41].

Author details

Takaki Iwata

The Institute of Statistical Mathematics, Tachikawa, Tokyo, Japan

References

[1] Woessner J, Wiemer S. Assessing the quality of earthquake catalogues: Estimating the magnitude of completeness and its uncertainty. Bulletin of the Seismological Society of America 2005;95(2):684-698.

[2] Iwata T. Low detection capability of global earthquakes after the occurrence of large earthquakes: investigation of the Harvard CMT catalogue. Geophysical Journal International 2008;174(3):849-856.

[3] Schorlemmer D, Woessner J. Probability of detecting an earthquake. Bulletin of the Seismological Society of America 2008;98(5):2103-2117.

[4] Schorlemmer D, Mele F, Marzocchi W. A completeness analysis of the National Seismic Network of Italy. Journal of Geophysical Research-Solid Earth 2010;115B4:B04308;doi10.1029/2008JB006097.

[5] Nanjo KZ, Ishibe T, Tsuruoka H, Schorlemmer D, Ishigaki Y, Hirata N. Analysis of the completeness magnitude and seismic network coverage of Japan. Bulletin of the Seismological Society of America 2010;100(6):3261-3268.

[6] Mignan A, Werner MJ, Wiemer S, Chen CC, Wu YM. Bayesian Estimation of the Spatially Varying Completeness Magnitude of Earthquake Catalogs. Bulletin of the Seismological Society of America 2011;101(3):1371-1385.

[7] Plenkers K, Schorlemmer D, Kwiatek G. On the Probability of Detecting Picoseismicity. Bulletin of the Seismological Society of America 2011;101(6):2579-2591.

[8] Shimshoni.M. Evidence for Higher Seismic Activity During Night. Geophysical Journal of the Royal Astronomical Society 1971;24(1):97-99.

[9] Flinn EA, Blandford.RR, Mack H. Evidence for Higher Seismic Activity During Night. Geophysical Journal of the Royal Astronomical Society 1972;28(3):307-309.

[10] Knopoff L, Gardner JK. Higher seismic activity during local night on the raw worldwide earthquake catalogue. Geophysical Journal of the Royal Astronomical Society 1972;283:311-313.

[11] Iwata, T. Quantitative analysis of the daily variation of earthquake detection capability. 2012;Chikyu Monthly;34(9):504-508 (in Japanese).

[12] Funasaki, J., Earthquake Prediction Information Division, Seismological and Volcanological Department, Japan Meteorological Agency. Revision of the JMA Velocity Magnitude. Quarterly Journal of Seismology 2004;67(1-4):11-20(in Japanese with English abstract and figure captions).

[13] Katsumata, A. (2004) Revision of the JMA Displacement Magnitude. Quarterly Journal of Seismology 2004;67(1-4):1-10(in Japanese with English abstract and figure captions).

[14] Nanjo KZ, Tsuruoka H, Hirata N, Jordan TH. Overview of the first earthquake forecast testing experiment in Japan. Earth Planets and Space 2011;63(3):159-169.

[15] Ogata Y, Katsura K. Analysis of Temporal and Spatial Heterogeneity of Magnitude Frequency-Distribution Inferred from Earthquake Catalogs. Geophysical Journal International 1993;113(3):727-738.

[16] Gutenberg B, Richter CF. Frequency of earthquakes in California. Bulletin of the Seismological Society of America 1944;34(4):185-188.

[17] Ringdal F. On the estimation of seismic detection thresholds. Bulletin of the Seismological Society of America 1975;656:1631–42.

[18] Iwata, T. Revisiting the global detection capability of earthquakes during the period immediately after a large earthquake: considering the influence of intermediate-depth and deep earthquakes. Research in Geophysics 2012;2(1):24–28.

[19] Powell MJD. Approximation theory and methods. New York: Cambridge University Press; 1981.

[20] Good IJ, Gaskins RA. Nonparametric Roughness Penalties for Probability Densities. Biometrika 1971;58(2):255-277.

[21] Eggermon PPB, LaRiccia VN. Maximum Penalized Likelihood Estimation, vol I: Density Estimation. New York: Springer; 2001.

[22] Good IJ. The estimation of probabilities. Cambridge: The MIT Press; 1965.

[23] Akaike H. Likelihood and Bayes Procedure. In:Bernardo JE. et al. (eds.) Bayesian Statistics. Valencia: University Press;1980. p143-166.

[24] Kass RE, Raftery AE. Bayes factors. Journal of the American Statistical Association 1995;90(430):773-795.

[25] Gill, P. E., Murray, W., Wright, MH. Numerical linear algebra and optimization, vol. I. Redwood City: Addison-Wesley; 1991.

[26] Tierney L, Kadane JB. Accurate Approximations for Posterior Moments and Marginal Densities. Journal of the American Statistical Association 1986;81(393):82-86.

[27] Konishi, S., Kitagawa, G. Information Criteria and Statistical Modeling Springer, New York; 2008.

[28] Atef AH, Liu KH, Gao SS. Apparent Weekly and Daily Earthquake Periodicities in the Western United States. Bulletin of the Seismological Society of America 2009;99(4):2273-2279.

[29] Iwata T, Young RP. Tidal stress/strain and the b-values of acoustic emissions at the Underground Research Laboratory, Canada. Pure and Applied Geophysics 2005;162(6-7):1291-1308.

[30] Nanjo KZ, Hirata, N, Obara, K., Kasahara, K. Decade-scale decrease in b value prior to the M9-class 2011 Tohoku and 2004 Sumatra quakes. Geophysical Research Letters 2012;39(20):L20304;doi:10.1029/2012GL052997.

[31] Scholz, C. The frequency-magnitude relation of microfracturing in rock and its relation to earthquakes 1968;58(1):399-415.

[32] Amitrano D. Brittle-ductile transition and associated seismicity: Experimental and numerical studies and relationship with the *b* value. Journal of Geophysical Research-Solid Earth 2003;108(B1):2044;doi:10.1029/2001JB000680.

[33] Lei XL. How do asperities fracture? An experimental study of unbroken asperities. Earth and Planetary Science Letters 2003;213(3-4):347-359.

[34] Mori J, Abercrombie RE. Depth dependence of earthquake frequency-magnitude distributions in California: Implications for rupture initiation. Journal of Geophysical Research-Solid Earth 1997;102(B7):15081-15090.

[35] Gerstenberger M, Wiemer S, Giardini D. A systematic test of the hypothesis that the b value varies with depth in California. Geophysical Research Letters 2001;28(1):57-60.

[36] Amorèse, D., Grasso JR, Rydelek PA. On varying *b*-values with depth: results from computer-intensive tests for Southern California. Geophysical Journal International 2010;180(1):347-360.

[37] Schorlemmer D, Wiemer S, Wyss M. Variations in earthquake-size distribution across different stress regimes. Nature 2005;437(7058):539-542.

[38] Narteau C, Byrdina S, Shebalin P, Schorlemmer D. Common dependence on stress for the two fundamental laws of statistical seismology. Nature 2009;462(7273):642-645.

[39] Fyen J. Diurnal and Seasonal-Variations in the Microseismic Noise-Level Observed at the Noress Array. Physics of the Earth and Planetary Interiors 1990;63(3-4):252-268.

[40] Hillers G, Ben-Zion Y. Seasonal variations of observed noise amplitudes at 2-18 Hz in southern California. Geophysical Journal International 2011;184(2):860-868.

[41] Wessel P., Smith, WHF. New, improved version of the Generic mapping Tools released. EOS 1998;79:579.

Permissions

The contributors of this book come from diverse backgrounds, making this book a truly international effort. This book will bring forth new frontiers with its revolutionizing research information and detailed analysis of the nascent developments around the world.

We would like to thank Sebastiano D'Amico, for lending his expertise to make the book truly unique. He has played a crucial role in the development of this book. Without his invaluable contribution this book wouldn't have been possible. He has made vital efforts to compile up to date information on the varied aspects of this subject to make this book a valuable addition to the collection of many professionals and students.

This book was conceptualized with the vision of imparting up-to-date information and advanced data in this field. To ensure the same, a matchless editorial board was set up. Every individual on the board went through rigorous rounds of assessment to prove their worth. After which they invested a large part of their time researching and compiling the most relevant data for our readers. Conferences and sessions were held from time to time between the editorial board and the contributing authors to present the data in the most comprehensible form. The editorial team has worked tirelessly to provide valuable and valid information to help people across the globe.

Every chapter published in this book has been scrutinized by our experts. Their significance has been extensively debated. The topics covered herein carry significant findings which will fuel the growth of the discipline. They may even be implemented as practical applications or may be referred to as a beginning point for another development. Chapters in this book were first published by InTech; hereby published with permission under the Creative Commons Attribution License or equivalent.

The editorial board has been involved in producing this book since its inception. They have spent rigorous hours researching and exploring the diverse topics which have resulted in the successful publishing of this book. They have passed on their knowledge of decades through this book. To expedite this challenging task, the publisher supported the team at every step. A small team of assistant editors was also appointed to further simplify the editing procedure and attain best results for the readers.

Our editorial team has been hand-picked from every corner of the world. Their multi-ethnicity adds dynamic inputs to the discussions which result in innovative outcomes. These outcomes are then further discussed with the researchers and contributors who give their valuable feedback and opinion regarding the same. The feedback is then collaborated with the researches and they are edited in a comprehensive manner to aid the understanding of the subject.

Apart from the editorial board, the designing team has also invested a significant amount of their time in understanding the subject and creating the most relevant covers. They scrutinized every image to scout for the most suitable representation of the subject and create an appropriate cover for the book.

The publishing team has been involved in this book since its early stages. They were actively engaged in every process, be it collecting the data, connecting with the contributors or procuring relevant information. The team has been an ardent support to the editorial, designing and production team. Their endless efforts to recruit the best for this project, has resulted in the accomplishment of this book. They are a veteran in the field of academics and their pool of knowledge is as vast as their experience in printing. Their expertise and guidance has proved useful at every step. Their uncompromising quality standards have made this book an exceptional effort. Their encouragement from time to time has been an inspiration for everyone.

The publisher and the editorial board hope that this book will prove to be a valuable piece of knowledge for researchers, students, practitioners and scholars across the globe.

List of Contributors

J.A. Peláez
Dpt. of Physics, University of Jaén, Jaén, Spain

J.C. Castillo
Dpt. of Historical Heritage, University of Jaén, Jaén, Spain

F. Gómez Cabeza
CAAI (Andalusian Center of Iberian Archaeology), Jaén, Spain

M. Sánchez Gómez
Dpt. of Geology, University of Jaén, Jaén, Spain

J.M. Martínez Solares
Section of Geophysics, IGN (National Geographical Institute), Madrid, Spain

C. López Casado
Dpt. of Theoretical Physics, University of Granada, Granada, Spain

M. Hamdache and A. Talbi
Seismological Survey Department, C.R.A.A.G. Algiers, Algeria

Septimius Mara
Ministry of Environment and Forests, Romania, Bucharest, Romania

Serban-Nicolae Vlad
Faculty of Ecology and Environmental Protection, The Ecological University, Romania

Mario Fernandez Arce
Escuela de Geología, Universidad de Costa Rica, Programa PREVENTEC, Red Sismológica Nacional (RSN: ICE-UCR). San José, Costa Rica, Central America

Zonghu Liao and Ze'ev Reches
School of Geology and Geophysics, University of Oklahoma, Norman, USA

Giuseppe Carlo Marano and Sara Sgobba
Department of Civil and Architectural Science, Technical University of Bari, Bari, Italy

Mariantonietta Morga
Mobility Department – Transportation Infrastructure Technologies, Austrian Institute of Technologies GmbH, Vienna, Austria

Sebastiano D'Amico
Department of Physics, University of Malta, Msida, Malta

Yoshiharu Shumuta
Civil Engineering Laboratory, Central Research Institute of Electric Power Industry, Chiba, Japan

Takaki Iwata
The Institute of Statistical Mathematics, Tachikawa, Tokyo, Japan

Printed in the USA
CPSIA information can be obtained
at www.ICGtesting.com
JSHW011408221024
72173JS00003B/464